2011
万达商业规划
WANDA COMMERCIAL PLANNING

中国建筑工业出版社

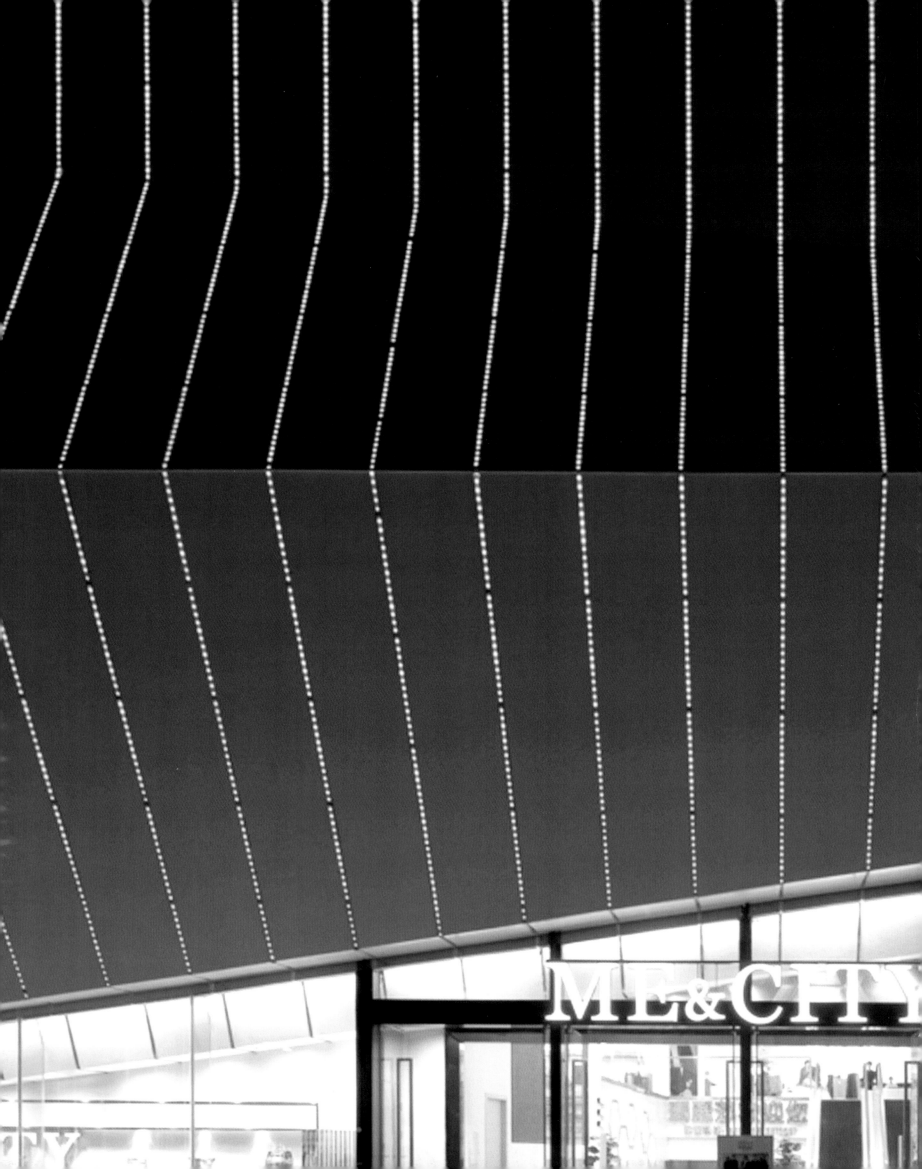

厦门湖里万达广场

常州新北万达广场室内步行街

《万达商业规划2011》编委会
THE EDITORIAL BOARD OF WANDA COMMERCIAL PLANNING 2011

主编单位
Chief Editor
万达商业规划研究院
Wanda Commercial Planning & Research Institute

规划总指导
General Advisor in Planning
王健林
Wang Jianlin

编委
Executive Editors
赖建燕　黄大卫　朱其玮　叶宇峰　冯腾飞　王元
Lai Jianyan, Huang Dawei, Zhu Qiwei, Ye Yufeng, Feng Tengfei, Wang Yuan

参编人员
Editors
马红　刘冰　孙培宇　莫力生　李峥　范珑　刘阳　万志斌　熊伟　王群华　侯卫华　王鑫
毛晓虎　郝宁克　袁志浩　阎红伟　黄引达　耿大治　黄勇　张振宇　王雪松　谢冕　高振江
孙辉　刘江　徐立军　刘杰　黄国辉　张立峰　章宇峰
Ma Hong, Liu Bing, Sun Peiyu, Mo Lisheng, Li Zheng, Fan Long, Liu Yang, Wan Zhibin, Xiong Wei, Wang Qunhua, Hou Weihua, Wang Xin, Mao Xiaohu, Hao Ningke, Yuan Zhihao, Yan Hongwei, Huang Yinda, Geng Dazhi, Huang Yong, Zhang Zhenyu, Wang Xuesong, Xie Mian, Gao Zhenjiang, Sun Hui, Liu Jiang, Xu Lijun, Liu Jie, Huang Guohui, Zhang Lifeng, Zhang Yufeng

李峻　梅咏　孙多斌　杨旭　杨成德　田杰　朱莹洁　李斌　门瑞冰　常宇　康军　吕永军　国文
谭耀辉　杨世杰
Li Jun, Mei Yong, Sun Duobin, Yang Xu, Yang Chengde, Tian Jie, Zhu Yingjie, Li Bin, Men Ruibin, Chang Yu, Kang Jun, Lu Yongjun, Guo Wen, Tan Yaohui, Yang Shijie

校对
Proofreaders
兰峻文　张涛
Lan Lunwen, Zhang Tao

英文翻译及校对
Translators and Proofreaders
吴昊　张震　沈文忠　梅林奇　宋锦华　刘佩
Wu Hao, Zhang Zhen, Shen Wenzhong, Mei Linqi, Song Jinhua, Liu Pei

WANDA **2011**

精益求精是品质提升的有效保障
REFINE ON IS AN EFFECTIVE GUARANTEE FOR QUALITY IMPROVEMENT

求精就要突出抓重点项目。项目公司要把项目的环境抓好,像绣花一样做好每个细节;对于泛光照明,规划院要精益求精。年底集团将由副总裁、万达学院院长主持,从工程、装饰、泛光照明、环境、招商等各个方面对项目进行评比打分,看谁做得最好。今后万达每年都要对各店进行评比,作为一项制度坚持下去。

Refinement is to highlight the grasp of key projects. The project company are required to maintain the site clean, to do a good job in every detail just like to do embroidery; The Institute is required to constantly strive for excellence. At the end of the year there will be a competition chaired by the Vice President of Wanda Group and the President of Wanda University, to select the best project from the aspects of project management, decoration, lighting, environment, merchant inviting and so on. This competition will persist as a system.

——王健林董事长,万达集团2011年半年会
Wang Jianlin
Chairman of Board and President of Wanda Group
At 2011 Semiannual Meeting

万达集团董事长
王健林
Wang Jianlin
Chairman of Board and President of Wanda Group

为全面提升万达广场的项目品质,由商业地产研究部、万达学院联合其他部门对今年开业的广场、酒店进行综合排名。这个排名不考察财务指标,只考察管理品质、内装、外观、招商情况、泛光照明这五条。排名的目的是拿第一名作为标杆,研究怎么整改。去年包河万达广场是今年的标杆,今年也要选一个做得最好的广场作为明年的标杆,促进万达广场品质不断提升。

Inorder to improve the quality of Wanda Plazas, our Commercial Project Research Department and Wanda Institute makes a ranking list of all the projects which opened this year. Financial indicators are not considered in this list, only five points are taken into account, which are management, interior decoration, outside appearance, rent and sale condition, and outside lighting. Our goal is to set the No. 1 as an example, and the others can learn from it. Baohe Wanda Plaza won the first prize of this year. We will choose another project which does the best job for the next year, to improve the quality of Wanda Plaza constantlty.

——王健林董事长,万达集团2011年半年会
Wang Jianlin
Chairman of Board and President of Wanda Group
At 2011 Semiannual Meeting

目录
CONTENTS

万达商业规划 2011	012	WANDA COMMERCIAL PLANNING 2011
万达广场设计管控	014	DESIGN MANAGEMENT AND CONTROL OF WANDA PLAZA
万达广场及大型购物中心消防与性能化论证	018	ON PERFORMANCE - BASED FIRE PROTECTION DESIGN OF WANDA PLAZA AND LARGE SHOPPING CENTERS
万达广场	023	PART 1 WANDA PLAZAS
福州仓山万达广场	024	FUZHOU CANGSHAN WANDA PLAZA
厦门湖里万达广场	036	XIAMEN HULI WANDA PLAZA
泰州万达广场	050	TAIZHOU WANDA PLAZA
唐山路南万达广场	060	TANGSHAN LUNAN WANDA PLAZA
常州新北万达广场	066	CHANGZHOU XINBEI WANDA PLAZA
银川金凤万达广场	076	YINCHUAN JINFENG WANDA PLAZA
镇江万达广场	084	ZHENJIANG WANDA PLAZA
武汉经开万达广场	088	WUHAN JINGKAI WANDA PLAZA
上海江桥万达广场	094	SHANGHAI JIANGQIAO WANDA PLAZA
廊坊万达广场	102	LANGFANG WANDA PLAZA
郑州中原万达广场	110	ZHENGZHOU ZHONGYUAN WANDA PLAZA
石家庄裕华万达广场	114	SHIJIAZHUANG YUHUA WANDA PLAZA
大庆萨尔图万达广场	120	DAQING SAERTU WANDA PLAZA
淮安万达广场	124	HUAI'AN WANDA PLAZA
酒店	129	PART 2 HOTELS
武汉万达威斯汀酒店	130	THE WESTIN WUHAN WUCHANG
广州白云万达希尔顿酒店	138	HILTON GUANGZHOU BAIYUN
济南万达凯悦酒店	142	HYATT REGENCY JI'NAN
南京万达希尔顿酒店	148	HILTON NANJING
唐山万达洲际酒店	154	INTERCONTINENTAL TANGSHAN
石家庄万达洲际酒店	158	INTERCONTINENTAL SHIJIAZHUANG
西安万达希尔顿酒店	164	HILTON XI'AN
常州万达喜来登酒店	170	SHERATON CHANGZHOU XINBEI HOTEL
泰州万达希尔顿逸林酒店	174	DOUBLETREE BY HILTON TAIZHOU, JIANGSU
镇江万达喜来登酒店	178	SHERATON ZHENJIANG HOTEL
廊坊万达希尔顿逸林酒店	184	DOUBLETREE BY HILTON LANGFANG
大庆万达喜来登酒店	190	SHERATON DAQING HOTEL
万达学院	193	PART 3 WANDA INSTITUTE
景观	207	PART 4 LANDSCAPE
项目索引	219	PART 5 INDEX OF THE PROJECTS
万达学院	222	WANDA INSTITUTE
万达商业规划研究院文化旅游分院	224	THE CULTURAL TOURISM BRANCH OF WANDA COMMERIAL PLANNING & RESEARCH INSTITUTE

万达商业规划2011
WANDA COMMERCIAL PLANNING 2011

万达集团董事长王健林指导规划院工作
Wang Jianlin guiding Wanda Commercial Planning & Research Institute

2011年是万达集团商业地产再次提速发展的一年。2011年开业的万达广场共16座,开业酒店12个,酒店的开业数量比2010年增加5个,是2009年(2个)的6倍。

The year of 2011 witnessed another accelerated growth of the commercial real estate of Wanda Group. 16 Wanda Plazas and 12 hotels were opened for business in 2011. Compared with 2010, the number of hotels opened for business increased by 5, while it is 6 times the number in 2009, namely 2.

2011年是万达商业规划研究院的品质年。

2011 is the year of quality of Wanda Commercial Planning & Research Institute.

2011年的万达商业规划研究院已初具规模,拥有200人,专业齐备。万达商业规划研究院总结了之前的项目设计及管控经验,使得2011年开业的项目在设计品质和最终呈现的建筑品质上较前几年开业的广场和酒店有了普遍的提升。2011年,万达商业规划研究院除了完成当年、次年开业项目的设计和管控外,对这两年的已设计完的开业项目及之前已建成开业的部分项目,从效果品质、招商品质等方面均进行了梳理和调整。从下面一组数据可以看出,万达商业规划研究院在提升设计品质上的巨大投入。

In 2011, Wanda Commercial Planning & Research Institute has reached a relatively large size, with 200 employees and a complete combination of different specialties. The Institute summarized the designing and controlling experience its previous projects of and materialized a general upgrading in the quality of designing and the final product, in comparison with Wanda Plazas and hotels opened in previous years. In 2011, apart from finishing the designing and controlling of projects to be opened in 2011 and 2012, the Institute also made some streamlining and adjustments on the quality of effects and investment promotion of the projects already designed in these two years as well as the some projects already opened previously. The set of data below can well demonstrate the huge effort by the Institute in upgrading the quality of its designing.

2011年,万达商业规划研究院组织完成了万达广场品质提升工作共5类108项。

In 2011, Wanda Commercial Planning & Research Institute organized and finished Wanda Plaza quality promotion work in 108 projects, which could be divided in 5 kinds.

1. 对已完成的20个商业项目的外立面进行重新调整及重新招标设计,使2011年及2012年开业的万达广场和酒店的形象更加时尚和个性化;

1. Readjusting, re-bidding and redesigning the facades of 20 commercial projects that have been built, to make the image of Wanda Plazas and hotels opened in 2011 and 2012 more fashionable and personalized;

2. 对已建成开业的29个项目的外立面店招、广告进行检查并整改，取消后期运营中增加的不当的广告店招；

2. Checking and rectifying the shop logos and advertisements on the facades of 29 projects that have been opened and cancelling improper shop logos and advertisements added in the operational period;

3. 对已建成开业的28个万达广场中的院线的普通影厅进行IMAX厅改造，提升万达广场影厅的科技含量和档次；

3. Renovating the IMAX halls of the common halls of cinemas in the 28 Wanda Plazas already built and opened, to upgrade the technical level and class of the cinemas of Wanda Plazas;

4. 对17个2012年及以后开业的万达广场的次主力店进行增减调整及布局修改，使业态比例更加合理；

4. Adjusting the number and layout of sub-anchor stores in 17 Wanda Plazas to be opened in and after 2012, to make the proportion of different formats more reasonable;

万达集团执行董事、总裁丁本锡对规划院工作进行现场指导
The executive director / president of Wanda Group Ding Benxi guiding Wanda Commercial Planning & Research Institute

5. 对14个2012年及以后开业的万达广场的娱乐楼独立门厅进行设计调整及布局修改，使流线更加适合运营管理。

5. Adjusting the designing and layout of the detached entrance halls of the entertainment buildings of the 14 Wanda Plazas opened in and after 2012, to make the flow more suitable for operation and management.

2011年，万达商业规划研究院除了完成万达广场的品质类工作的梳理外，还对23个万达广场周边的室外步行街的动线进行了总图优化，对已开业的11个室外步行街进行了景观、夜景照明、美陈等商业氛围营造性改造；同时也对13个在设计中的酒店的标准层衣帽间进行步入式设计调整，使万达的高星级酒店客房更加奢华。

In 2011, apart from streamlining the jobs concerning the quality of Wanda Plazas, the Institute also optimized the master planning of the circulation of the outdoor pedestrian street around 23 Wanda Plazas and conducted constructive renovations on the 11 outdoor pedestrian street already opened, in terms of their commercial atmosphere including landscaping, nightscape lighting and art display. At the same time, the adjustment was also made on the walk-in design of the closets on the standard floors of 13 hotels being designed, to make Wanda's high-star-level hotels more luxurious.

2011年，大连万达商业地产股份有限公司首次对当年开业的万达广场进行了"品质"评比。从"设计品质"、"建造品质"、"招商品质"三个方面，以及建筑空间形象、内装、景观、夜景照片、室内店铺的一店一色等效果和动线功能，室内舒适度等合理性进行全面的品质评估。当年，福州仓山万达广场、厦门湖里万达广场和泰州万达广场被评为优秀商业项目。2011年集团对品质的重视为此后万达广场及酒店建设的品质提升奠定了基础。

In 2011, Dalian Wanda Commercial Estate Co.,LTD. firstly had a competition on the quality of new-opened Wanda plazas, and had a comprehensive assessment form three aspect of "quality of design", "quality of construction", "quality of merchant inviting", as well as its architecture space, decoration, landscape, nightscape, color of the indoor-shop and function. At that year, Fuzhou Cangshan Wanda Plaza, Xiamen Huli Wanda Plaza and Taizhou Wanda Plaza were titled excellent commercial programme. In 2011, paying attention to the construction quality provided the basis for the excellence of Wanda Plaza and Hotel.

大连万达商业地产股份有限公司高级总裁助理
万达商业规划研究院院长
赖建燕
Lai Jianyan
Senior Assistant of the President of
Dalian Wanda Commercial Estate Co., LTD
President of Wanda Commercial Planning & Research Institute

万达广场设计管控
DESIGN MANAGEMENT AND CONTROL OF WANDA PLAZA

在万达广场的开发建设过程中，设计管控自始至终贯穿整个工程项目的全过程。它不同于传统意义上的设计单位对设计的管理，而是从项目选址、概念规划、方案设计、初步设计、施工图设计、材料封样、规划验收、运营整改、复盘总结和成果转换等方面全周期的过程管控，对项目成本的控制、安全的保证以及整个商业地产未来的运作有着十分重要的意义。

Design management and control runs through the whole process of developing and constructing of Wanda Plaza. Unlike the conventional management of design by the designer, our management and control covers the complete progression from site selection, conception plan, schematic design, preliminary design, drawing design, material specifications, planning acceptance, operation rectification, and review and summary through to result conversion. It is of great significance to project cost control, safety guarantee and future operations of the entire commercial properties.

1 设计管控的内容
1 The Content of Design Management and Control

设计管控分为阶段管控、重点管控和专项管控三个方面。

Design Management and Control is divided into 3 aspects, namely stage management and control, management and control priorities, and special project management and control.

1.1 阶段管控
1.1 Stage Management and Control

设计管控按阶段划分为概念规划、方案设计、初步设计、施工图、项目实施、运营六个阶段，在不同的阶段有不同的管控重点。

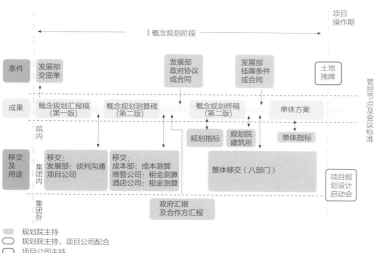

图1: 概念规划阶段设计管控流程图
Figure 1 Flow Chart of Design Management and Control in the stage of Conception Planning

According to different stages, design management and control is divided into 6 stages, namely concept plan, schematic design, preliminary design, architectural drawing, project implementation, and operation, each underling a key aspect.

在概念规划阶段，除了规划设计本身以外，配合前期发展部门进行谈判、决策和成本管理部门进行成本测算是这个阶段管控的重点。

The focus of the concept plan stage is to support development departments in negotiation and making decision and to assist cost control departments in cost estimation.

方案设计阶段则主要侧重于项目的设计单位招标工作和方案评审、方案论证以及对设计单位的方案阶段成果进行检查；初步设计阶段，对空调冷热源、供电方案等专项进行论证是管控的重点；施工图设计阶段的管控则侧重于消防报审、报批报建、与招商部门的对接等方面。

The schematic design phase emphasizes bid invitation for the project design, schematic evaluation, schematic demonstration and inspection of the progress made by designer at different stages. The preliminary design phase underlines the demonstration of the cooling and heating sources of air-conditioning and power supply scheme. The architectural drawing design phase stresses requesting approval for fire control, submitting for permission and construction and matchmaking activities with investment promotion departments.

到了项目实施阶段和运营阶段，主要的管控点是对实施情况的检查和规划验收，要求负责实施的部门对项目实施和运营过程中出现的问题进行整改、复盘，形成合理有效的成果，用以指导以后项目的设计工作。

The project implementation and operation phases focus on supervision of the implementation, planning inspection and acceptance, which ask the competent departments to rectify and review problems occurring during the project implementation and operation, with a view to achieving optimal design effect and guiding the future project design.

1.2 重点管控

1.2 Management and Control Priorities

计划控制、定额设计、面积管控、项目品质、安全质量、节能绿建和强条落实七大方面是万达广场的重点管控内容。

Management and control priorities of Wanda Plaza cover 7 aspects, namely plan control, quota design, area management and control, project quality, safety, conservation and green building, and regulation enforcement.

计划管控是万达广场购物中心得以顺利建设的重要保障，它通过计划模块的管理体系实现，各部门遵照规定的计划模块开展工作，保证模块中每一个节点的顺利完成。规划设计计划是计划管控的重要组成部分，涉及节点数92个，约占总计划模块的30%，具有周期最长、关联最广的特点。

Plan control, an important guarantee of the smooth construction of Wanda Plaza shopping center, is realized through the management system of plan modules. All departments act on the prescribed plan modules, securing a smooth run of every single node of plan modules. Planning of schematic design, a major part of plan management and control, touches upon 92 nodes which account for 30% of the overall plan modules, and has the longest period and the most extensive correlations.

定额设计是商业地产快速发展的必然产物，它是在满足项目品质和安全的前提下，对设计成本管控的标准化体现。定额设计以建造标准为基础，对构造做法、节点大样、材料选样、设备选型等进行优化、量化、细化，并一一对应成本指标形成定额设计标准图集、做法表等，达到从设计阶段开始即进行成本管控的目的。

Quota design is the inevitable product of the rapid development of commercial real estate. It is the manifestation of the standardization design cost control, with the quality and safety of the project as the premise. Quota design is based on the construction standards, so the structure, the details, material sampling, and equipment selection can be optimized, quantificational and refinement, and all of these can be issued to the quota design standard tables, schedules, etc., and get the target of cost control from design phase.

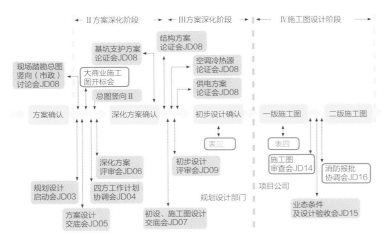

图2: 方案、初设、施工图阶段设计管控流程图
Figure 2 Flow Chart of Design Management and Control in the stages of Schematic Plan, Preliminary Design and Drawing Design

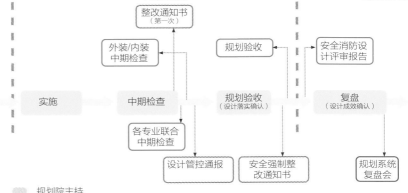

图3: 实施、运营阶段设计管控流程图
Figure 3 Flow Chart of Design Management and Control in the stages of Implementation and Operation

定额设计涵盖持有物业和销售物业所有产品类型，涉及全专业内容。一键式集中控制设计标准及运行管理标准的建立，将商业运营阶段也纳入定额设计的范畴，从而使项目从前期设计到建成运营全阶段都在定额设计范围内开展工作，真正达到成本管控的目的。

Quota design covers all product types, self-hold property and sale property, which is involved in the all-professional content. The setup of centralized management of design standard and standards of operation and management, will include the commercial operation phase into the scope of quota design, in order to really achieve the purpose of cost control with the project being working under the quota design from preliminary design phase to the operation phase.

万达广场从项目前期到建成运营都应在定额设计的要求下开展工作以控制投资成本的增加。面积管控的目的是为决策层提供及时、准确、高效的数据支持，并保证设计成果的落实和连贯性。

In order to control the investment cost, all the work shall be done in light with the requirement of quota design from the initial stage of constructing to its operation. The purpose of area control is to provide decision-makers with timely, accurate and efficient data support, and ensure the implementation and coherence of the design.

设计管控的核心就是为了项目的品质得到有效的保障，项目品质主要以考核的方式进行评判，考核的要点是总体效果和技术要点，其中总体效果包括：外立面、内装、夜景照明、景观、停车场等，技术要点包括：平面功能动线、外立面、内装、夜景照明、停车场、舒适度等。

The core of design management and control is to effectively ensure the project quality which is largely evaluated through assessment. The assessment is mainly on overall effect, including facade, interior, nightscape lighting, landscaping, parking lot, etc., and essential technologies, which cover plane circulation, facade, interior, nightscape lighting, parking lot, comfort, etc..

安全管控由集团安全管理制度和安全管控体系来实现从设计、建造到运营阶段的全程安全管理；质量管控是通过质量管理组织体系和质量管理制度的制订，实现保证质量的目的。

Security management and control is to make the criteria of acceptance and to regulate in accordance with it. Quality management and control is to establish the quality management system and institutions and ensure product quality.

万达集团对绿建节能工作一直高度重视,"绿色低碳"是万达一贯倡导的社会责任的重要方面,它是万达集团社会责任的体现。节能绿建以绿色建筑设计标识认证和运营标识认证作为主要考核目标。

Wanda Group highly values green building and energy conservation. The pursuit of the "Green and Low Carbon" principle has all along been an important part of, and a reflection of Wanda's social responsibilities. In respect of green building and energy conservation, the main objectives for assessment are to obtain the design certificate and the operation identity certificate of green buildings.

1.3 专项管控
1.3 Special Project Management and Control

对于重点或特殊类项目,通过打破集团内外各部门、打破集团内各系统、打破各专业之间的界限,合理配置更多、更高效的资源,仅就特殊专项进行管控,是行之有效的管控手段。专项管控由规划院牵头组织专项小组进行设计管控。

Special project management and control is an effective way to manage and control important and special projects by means of breaking down the barriers between different departments inside and outside the Group and between different systems and professions inside the Group, as well as reasonably allocating more resources with higher efficiency. Special project management and control is carried out by working groups that are organized and led by the Commercial Planning and Research Institute.

Design management and control covers the whole chain of commercial real estate development and is built on the general information platform of the Group, with the intelligent system of computer information as its basis. There are mainly four modes of it, i.e. planning informatization, informatization of the filling and submission of property area data, project & cost informatization, and fire & safety informatization.

设计管控通过外立面、室内步行街内装、景观、夜景照明、导向标识、弱电智能化这六类专项的细分,将整个设计管控过程化整为零,进行精细化管理。

Design management and control perfectly breaks down the whole process into six special projects, namely facade, interior decoration of indoor pedestrian street, landscaping, nightscape lighting, guiding signs and weak current intellectualization, and conducts delicacy management of each.

设计管控的依据是国家规范、行业标准和万达集团企业制度标准等。

The design management and control is based on national specifications, professional standards and Wanda Group's institutions and criteria, etc..

图4: 327个计划管控模块节点
Figure 4 327 Scheduled models

2 设计管控模式和依据
2 Modes and Bases of Design Management and Control

设计管控是在集团整体的信息化平台之下,以计算机信息化智能系统为基础,纳入整个商业地产开发链条的管控,包括四个重要的管控模式:计划模块的信息化,物业面积填报的信息化,规划与成本的信息化和消防安全的信息化。

3 设计管控与招商对接
3 Design Control and Management and Investment Matchmaking

设计管控与招商对接是万达广场购物中心建造、运营的前置条件,只有良好的招商对接才能实现其投资、建造的目的。

The connection between design control and management and tenant invitation is the prerequisite for constructing and running Wanda Plaza. Only with an effective connection to tenant invitation, can the goal of investment and construction be realized.

万达广场购物中心设计与招商的关系是"业态规划在先,建筑设计在后",是万达独创的"订单地产"的招商模式,能够确保项目建成之日,同期满铺开业。

The relationship between the design and tenant invitation of Wanda Plaza follows the principle of planning of business layout coming ahead of architectural design. This is a unique investment promotion pattern developed by Wanda. Such a mode can ensure that all shops of the same period open on the day the project is completed.

图 5：万达集团董事长王健林指导规划院工作
Figure 5　Wang Jianlin guiding Wanda Commercial Planning & Research Institute

万达集团由集团总部和项目公司两级系统构成，集团总部负责整合集团资源、制定统一标准，完成决策工作；地方项目公司结合实际情况，负责落实集团意图，快速实施建造。相关部门的工作关系（见图6）。

Wanda Group is a two-tier system, consisting of corporation companies and project companies. Corporation companies are in charge of the integration of group resources, the formulation of consistent standards, and decision-making, while local project companies are responsible for implementing the intention of the group and carrying out rapid construction in light of the local conditions. Please refer to Figure 6 for the work relationship among relevant departments.

随着万达广场第三代购物中心的实施，招商对接已按流水线的方式在项目建造全程流畅运作，所有项目所有关键节点均按固定计划表及对应节点标准实施、管控。商家与招商部门的配合已经炉火纯青，在对接过程中，综合成本、建造时间、政府部门审批、效果等多方面因素考虑，最终设计成果移交项目公司便可以顺利实施。

With the third-generation plaza well under way, the connected process of tenant invitation runs smoothly in an assembly-line fashion during the whole construction process. The key nodes of all projects are carried out, controlled and managed in compliance with the fixed schedule and their corresponding standards. The cooperation between businesses and tenant invitation departments has been perfect. All kinds of factors like comprehensive costs, construction time, approvals of government agencies and effects will be taken into consideration in the connection process. After handing over the final design to the project company, the tenant invitation can be carried out smoothly.

设计管控作为贯穿在整个工程开发过程的管理动作，其影响力之大、覆盖面之广，是有效把控项目最终效果、有效控制项目成本、保证项目按计划实施、保证招商顺利实施的重要手段。万达广场的建设正是依靠有效的设计管控，才能够保证每年二十个左右的项目顺利开业，不论是从数量上还是品质上均是商业建筑领域内的高端产品。

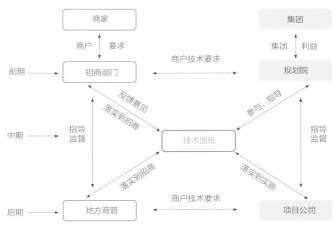

图 6：设计管控与招商对接关系示意图
Figure 6　Relationship between Design Control and Management and Investment Matchmaking

With its profound influence and extensive coverage, the design control and management as a procedure running through the whole developing process is an essential means to effectively bring the outcome and cost of the project under control and to ensure that the project is constructed as planned and tenant invitation goes well. It is with the help of effective design control and management that Wanda is able to successfully open some twenty projects every year. Wanda Plazas are among the high-end products in the field of commercial architecture, in terms of both quantity and quality.

万达商业规划研究院副院长
冯腾飞
Feng Tengfei
Vice President of Wanda Commercial Planning & Research Institute

万达广场及大型购物中心消防与性能化论证
ON PERFORMANCE - BASED FIRE PROTECTION DESIGN OF WANDA PLAZA AND LARGE SHOPPING CENTERS

随着中国城市化进程的发展，大型商业购物中心建设也进入高速发展阶段。集购物、娱乐、餐饮于一身的大型现代化购物中心，对改善城市面貌和人民居住生活条件起到了积极作用。万达广场就是现代购物中心非常典型的例子。万达广场购物中心建筑规模约15万平方米，长度300米左右，宽度将近160米，经营业态涵盖百货、零售、超市、娱乐等综合业态，商业空间通透、高大，为消费者提供了极其舒适的购物环境和消费体验。

万达广场室内步行街实景照片
A View of the Indoor Pedestrian Street in Wanda Plaza

With the development of urbanization in China, big commercial shopping centers have seen a rapid growth. Integrating shopping, recreation and dining, these large-scale modern shopping centers play an active role in improving city's image and living conditions. A typical example in this respect is Wanda Plaza. Covering an area of 150,000 square meters, its 300-meter-long and 160-meter-wide shopping center covers a wide range of business types including department stores, retailers, supermarkets, recreational venues, etc,. The bright and spacious commercial space creates a comfortable shopping environment and consumption experience for customers.

综合类大型商业建筑规模大、人流量大、内部空间复杂的特点，使其消防设计与普通民用建筑有明显区别。如发生火灾，造成的财产损失和人员伤亡更为严重。如何确保大型购物中心的消防安全，给消防设计工作提出了新的课题。

The great size, heavy pedestrian flow and intricate interior space of large comprehensive commercial buildings necessitate a fire protection design completely different from that of ordinary civil buildings. In case of fire, it would result in greater economic losses and casualties. Therefore how to secure large shopping centers from fire becomes a fresh question for fire protection design.

相对复杂体量建筑的不断建成投入使用，我国适用于此的消防规范的编制发行却显得相对滞后。《建筑设计防火规范》（GB 50016—2006）、《高层民用建筑设计防火规范》（GB 50045—2005）、《商店建筑设计规范》（JGJ 48—88）、《建筑内部装修设计防火规范》（GB 50222—95，2001年修订版）、《自动喷水灭火系统设计规范》（GB 50084—2005）……这些规范中的内容显然无法针对现在大型购物中心的实际情况给出明确措施和做法。那么如何在既满足商业经营需要的前提下，又保证消防设计的安全呢？

While large buildings with complicated interior designs keep cropping up in great numbers, the promulgation of applicable fire protection regulations seems lagging behind in China: *Code for Fire Protection Design of Buildings* (GB 50016-2006); *Code for Fire Protection Design of Tall Buildings* (GB 50045—2005); *Code for Design of Store Buildings* (JGJ48—88); *Code for Fire Prevention Design of Interior Decoration of Buildings* (GB 50222—95, revised in 2001); *Code of Design for Automatic Sprinkler Systems* (GB50084—2005)… Apparently, these codes did not provide for any specific measures or instructions tailored to the actual conditions of large shopping centers. Then how could we ensure a safe fire protection design while meeting the needs of business operation?

针对此种矛盾，发达国家率先开始了"性能化"防火设计的研究。自20世纪80年代英国提出了"以性能为基础的消防安全设计方法"的概念以来，截至目前，已有不少于13个国家（澳大利亚、加拿大、芬兰、法国、英国、日本、荷兰、新西兰、挪威、波兰、西班牙、瑞典和美国）正式颁布规范，采用或积极发展性能化规范和基于规范结构形式下的建筑防火设计方法，并取得了一定成果。

In response to this, developed countries took the lead in studying the "performance-based" fire protection design. Since 1980s when Britain put forward the concept of "performance-based fire protection design methods", at least 13 countries (Australia, Canada, Finland, France, Britain, Japan, Netherlands, New Zealand, Norway, Poland, Spain, Sweden and America) have issued official codes and garnered some achievements by applying or actively developing performance-based codes and fire protection designs based on normative structures.

性能化消防设计是以火灾安全工程学的思想为指导，建立以火灾性能为基础的建筑防火设计（Performance-based Fire Protection Design），在国内通常简称为"性能化"防火设计。性能化防火设计包括确立消防安全目标，建立可量化的性能要求，分析建筑物及内部情况，设定性能设计指标，建立火灾场景和设计火灾，对设计方案进行安全评估。在设计过程中，对建筑物可能发生的火灾进行量化分析，并对典型火灾场景下火灾及烟气的发展蔓延过程进行模拟计算，一般需要采用计算机火灾模拟软件等分析和计算工具。

Guided by the ideas of fire protection engineering, performance-based fire protection design is a fire protection solution that is based on the fire performance. In China, it is also known as "performance-based" fireproof design. The design involves setting up fire prevention goal, establishing quantifiable performance requirements, analyzing buildings and their interior, defining performance design indicators, projecting fire scenario and conjecturing fire, and making safety appraisal of the design. In the development of a fire protection strategy, analytic and computing tools are usually employed to make a qualitative analysis of the hypothetical fires that might happen to the buildings and make an analogy computation of how the flame and smoke would spread in a typical fire scenario.

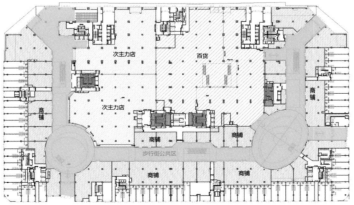

万达广场室内步行街安全区示意图
Illustration of Indoor Pedestrian Street Safety Area in Wanda Plaza

对于大型购物中心而言，消防性能化设计的核心点在于通过各种消防安全手段将共享空间作为"室内安全区"进行设计，实现对消防疏散的有效组织。制订整体消防策略，从而做出消防安全评估，以达到与依照传统建筑设计防火规范提出的消防策略相同的消防安全水平。同时，消防性能化设计对业主日后的消防管理提出了非常严格的要求。万达广场的消防设计率先开始性能化设计的探索。

Regarding to large shopping centers, the essence of performance fire protection design is to design the communal space as the "indoor safety area" by all means of fire suppression measures so as to efficiently organize the evacuation in the event of fire. An overall fire protection strategy needs to be developed so as to make a safety appraisal of fire prevention and achieve the same effect as the fire prevention strategy required by codes for fire protection design of traditional buildings. Moreover, performance fire protection design entails very strict fire safety management from building owners. Wanda Plaza is at the forefront of exploring performance fire protection design in China.

万达广场在消防性能化设计中，通过以下措施保证"室内安全区"成立。

Regarding the performance fire protection design of Wanda Plaza, the following measures have been taken to ensure the establishment of "indoor safety area".

1 划分防火单元，降低街两侧商铺火灾危险
1 Designate Fire Protection Units and Reduce the Risk of Fire in the Stores Lining the Street

针对室内步行街的使用和消防特点，引入"防火单元"的理念，将公共区域两侧火灾载荷大、用电设备多、火灾危险性高的每个小商铺作为一个"防火单元"，商铺之间、商铺与公共区域之间采用防火分隔，商铺内设置火灾报警、自动喷水灭火和机械排烟系统，确保起火情况下能迅速被扑灭，不造成火灾烟气的蔓延。

Taking into account the use and the fire prevention of indoor pedestrian street, we introduced the concept of "fire protection units", which are small stores lining the street. Because of the greater fire loads and multiple electrical equipments, those small stores face the high risk of fire. Therefore, discrete fire zones are created among stores and between stores and public areas by means of fire compartmentalization. Inside stores are installed fire alarms, automatic sprinkler system and mechanical smoke exhaust system, ensuring that fire could be put out immediately and smoke could be contained from spreading.

2 确保防火分隔界面完整有效
2 Ensure that the Fire Compartmentalization is Complete and Effective

商铺与公共区域之间的防火分隔界面设计对"室内安全区"的成立至关重要。万达商业规划研究院与国家消防工程技术研究中心一起进行的"室内步行街两侧商铺防火分隔技术研究",采用钢化玻璃加水喷淋冷却保护的方式,成功通过现场模拟实验的数据报告,满足了耐火极限的要求,很好地解决了这一问题。

The design of the fire compartmentalization between stores and public areas holds the key to the establishment of "indoor safety areas". The "Technical Research of Fire Compartmentalization in Stores Lining Indoor Pedestrian Street" coauthored by Wanda Commercial Planning and Research Institute and National Engineering Research Center for Fire Protection, a data report made on the basis of successful field simulation testing, provides a perfect solution to the problem by using toughened glass plus water spray cooling system which reaches the limit of fire resistance.

现场模拟实验
Field Simulation Test

3 多种消防措施到位确保"室内安全区"的成立
3 Put in Place Diverse Fire Suppression Measures to Ensure the Establishment of "Indoor Safety Areas"

通过对步行街公共区地面及墙面装修材料的级别、建筑构件的耐火极限及疏散楼梯布置原则、间距等的约定,并通过对室内步行街排烟系统、消防喷淋系统及智能疏散系统的设置等要求,确保火灾状况下室内步行街的安全性能,保证火灾时人员的安全疏散。

The safety of indoor pedestrian street must be guaranteed and people must be safely evacuated in the event of fire by prescribing the grade of the finishing materials used in the pedestrian street flooring and walls, the limit of fire resistance of building components, layout principles and spacing of evacuation stairs as well as by requesting the installation of smoke exhaust system, fire protection sprinkler system and intelligent evacuation system in the pedestrian street.

4 加强管理措施,提高消防标准
4 Strengthen Management Measures so as to Improve the Fire Protection Standards

万达广场购物中心由万达商业管理公司统一管理,通过管理措施确保消防安全,包括在建筑的所有出入口设置了客流计数系统;对电气竖井等相关设计内容提出严格的限制;在企业标准中增加了救援作业场地和消防救援窗等设置要求;对涉及消防的建筑产品建立专门的品牌库等针对性措施。

Under the unified management of Wanda Commercial Management Company, Wanda Plaza shopping centers are protected by a host of fire prevention measures. For instance, all the entrances and exits of buildings are installed with a customer counting system; the design of electric shafts must follow rigorous rules; in terms of enterprise standards, fresh requirements are added such as setting up sites for rescue operations and windows for fire rescue; and some targeted measures like creating a specialized brand database for building products related to fire prevention have been implemented.

近年来,数十座万达广场成功开业的事实充分证明,万达集团在大型购物中心消防领域开展的研究及应用,在消防安全设计及运营方面发挥了重要作用,也得到了各地消防部门的认可。

In recent years, dozens of Wanda Plaza have come into operation successfully, fully showing that Wanda Group has played a key role in studying and applying fire prevention technologies to large shopping centers as well as in fire protection design and operation, and its efforts has been recognized by fire fighting departments across the country.

万达集团总结多年的探索经验,由规划院牵头,联合国家消防工程技术研究中心,开展大型综合性商业建筑防火设计关键技术研究,先后编制出版了《万达广场购物中心建筑防火技术要求》等多项企业标准,并完成了《大型综合性商业建筑火灾荷载调查研究》、《大型综合性商业建筑人员荷载调查研究》、《室内步行街两侧商铺防火分隔技术研究》、《步行街中庭排烟系统及设计参数研究》、《首层步行街公共区作为人员疏散安全区的技术条件研究》5个专题研究报告,填补了国家消防理论的空白;并在以上研究成果基础之上完成的《大型综合性商业建筑防火设计关键技术研究》,已成为公安部应用创新科研项目,待最终审批通过后,将成为大型购物中心消防安全及性能化方面第一个权威性科研成果。

Referring to its exploration experience over the years and following the lead of Wanda Commercial Planning and Research Institute, Wanda Group, in joint efforts with National Engineering Research Center for Fire Protection, has launched studies of key fire protection technologies in large comprehensive commercial buildings, issued several corporate standards including Building Fire Protection Requirements for Wanda Plaza Shopping Center, and to date completed five special research reports, namely, Survey Research of Fire Load of Large Comprehensive Commercial Buildings, Survey Research of People Load of Large Comprehensive Commercial Buildings, Study of the Smoke Exhaust System and Design Parameters of Pedestrian Street Atrium, and Study of the Technological Qualifications to Make the Top Floor of Pedestrian Street Public Area as the Safe Zone for Evacuating People. Based on the above-mentioned studies, the Study of Key Fire Protection Technologies in Large Comprehensive Commercial Buildings was accomplished and counted as a scientific project of application innovation by the Ministry of Public Security. Once approved, it will be the first authoritative scientific work in the field of fire protection and performance-based fire protection design of large shopping centers.

万达商业规划研究院副院长
马红
Ma Hong
Vice President of Wanda Commercial Planning & Research Institute

PART 1　万达广场
WANDA PLAZAS

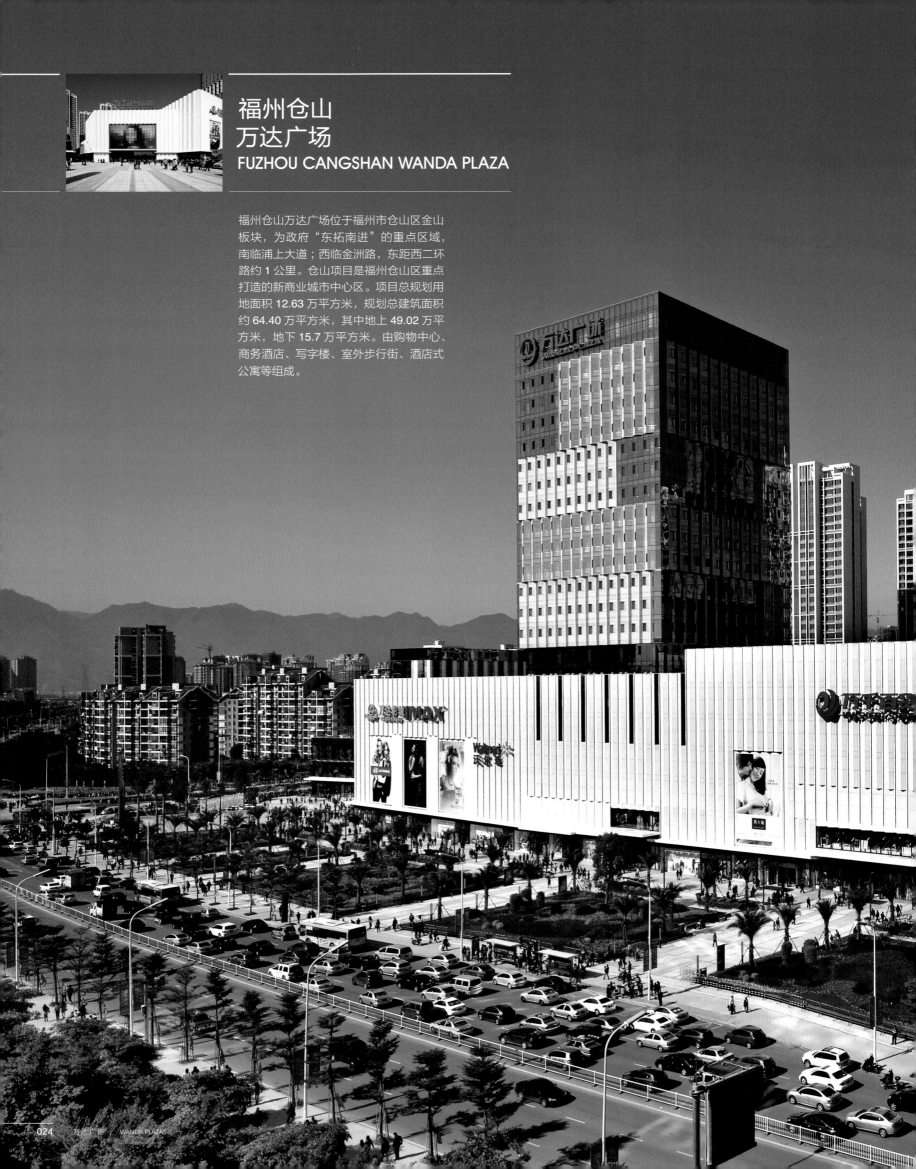

福州仓山
万达广场
FUZHOU CANGSHAN WANDA PLAZA

福州仓山万达广场位于福州市仓山区金山板块，为政府"东拓南进"的重点区域，南临浦上大道；西临金洲路，东距西二环路约1公里。仓山项目是福州仓山区重点打造的新商业城市中心区。项目总规划用地面积12.63万平方米，规划总建筑面积约64.40万平方米，其中地上49.02万平方米，地下15.7万平方米。由购物中心、商务酒店、写字楼、室外步行街、酒店式公寓等组成。

万达广场立面日景

Fuzhou Cangshan Wanda Plaza locates in the Jinshan area of Fuzhou Cangshan District, within the key area of government's "Eastern and Southenn exploitation strategy". The project faces Pushang Avenue on south and Jingzhou Road on west, and it is around 1km away from west No.2 Ring Road. Cangshan project is the important urban commercial center of Cangshan District. The site area is 126,300 m² and total building area is around 644,000 m², including 490,200 m² above ground and 157,000 m² in basement. Cangshan Plaza consists of shopping center, business hotel, office towers, outdoor promenade, service apartment buildings and so on.

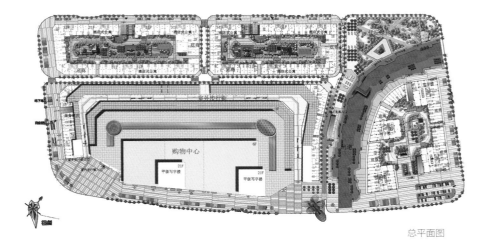

总平面图

建筑设计结合福建地域多山及临海的特点,引入了层岩与海洋的概念,
营造出德式建筑的大气、精细、俊朗,同时不失商业建筑的时尚与活跃

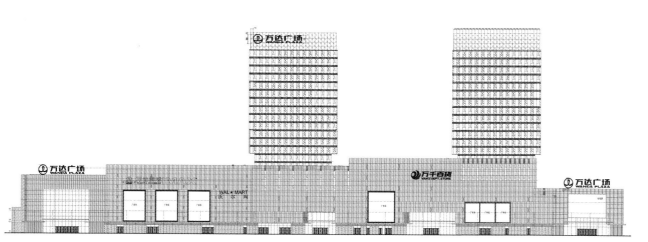

综合楼南立面剖面

建筑夜景

购物中心外立面整体采用波浪形金属幕墙，完美实现了层岩概念与海洋概念的结合，凸显福州地域特色，总体形象大气，细节考究，体现了高品质商业建筑应有的建筑形象。

The shopping center applies corrugated metal curtain wall system to the whole façade, which indicates the perfect integration of layered rock and ocean concepts, and highlights the Fuzhou regional characteristics. The overall project design expresses an imposing but elegant detailing, and high-quality commercial architecture imagery.

外广场设计营造了大片绿植及随处可见的休闲座椅，并充分利用水面资源，将河道两岸设计为层次丰富、尺度宜人的园林式环境，增加了商业建筑的人文气质，提升了万达广场的人性化与品位。福州仓山万达广场于 2010 年 7 月开工，于 2011 年 12 月开业。

The outer plaza includes a lot of planting and outdoor seating facilities. It takes full advantage of water body and converts the current riverside to be a multiple-layer, human-scale and garden-style promenade, upgrading the commercial building's humanity quality as well as the plaza's amenity and taste. Fuzhou Cangshan Wanda Plaza started construction in July, 2010 and opened in December, 2011.

精心设计的橱窗和外街商铺展示面，让人流连忘返，体现了高品质商业建筑应有的建筑形象

主立面入口夜景

主立面细部夜景

室外步行街

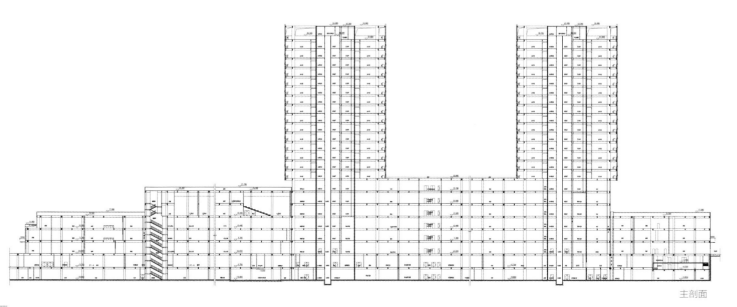

主剖面

波浪形外墙面完美实现了层岩概念与海洋概念的结合,凸显福州地域特色

广场外立面局部

广场外立面局部

利用中庭侧裙的形体起伏以及长街跳跃变化的灯光来表达仓山靠海的地域特点

万达影城

大玩家超乐场

大玩家超乐场

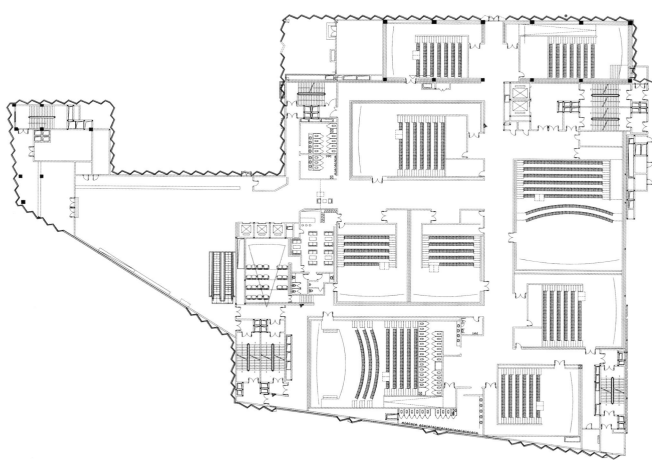
万达影城平面图

室内步行街入口

1F

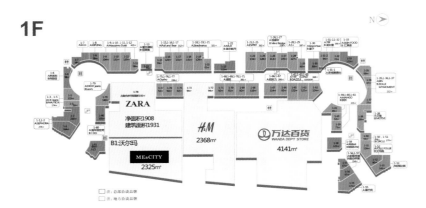

2F

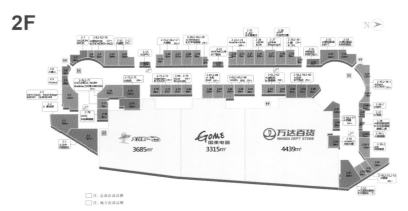

3F

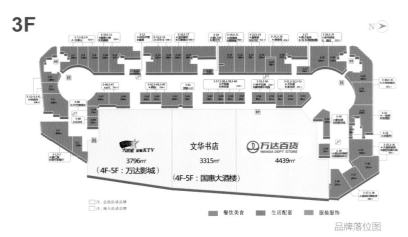

品牌落位图

室内步行街内装

厦门湖里
万达广场
XIAMEN HULI WANDA PLAZA

厦门湖里万达广场坐落于厦门岛的东北角，为城市新区地段，周边主要规划为高档住宅区，地块的东西南北均有城市干道，交通便捷，可快速方便地到达城市其他功能区。项目总规划用地面积 12.92 万平方米，规划总建筑面积约 52.92 万平方米，其中地上 38.82 万平方米，地下 14.10 万平方米。项目主要由购物中心、SOHO、室内步行街等构成。

广场夜景照明

总平面图

Xiamen Huli Wanda Plaza is located at the northeast corner of Xiamen Island in the new urban area. Its surroundings are planned to be high-end residential district. There are urban arterial roads on all directions of the plot with convenient transportation. It is rapid and convenient to reach other functional districts in the city. The overall site area is 129,200 m². The gross floor area is 529,200 m² with 388,200 m² above-ground and 141,000 m² under-ground area. This project mainly consists of shopping mall, SOHO, five-star hotel, and indoor pedestrian street.

购物中心钻石形的外观几何感极强,外墙面以白色及深灰色为主调,几何切割出的白色形体与独特的肋状外墙组合,使人过目难忘;环抱购物中心的是高层SOHO公寓,其立面以海洋文化为主题,采用波浪形装饰外挑板,共同打造出地标性极强的立面形态。同时,作为万达的第三代城市综合体,它以时尚、休闲、娱乐的业态定位,融合购物、娱乐及餐饮的一站式消费模式为厦门人带来崭新的消费体验。厦门湖里万达广场于2010年4月开工,于2011年9月开业。

The diamond-shaped appearance of the shopping mall endows strong geometric feelings. The exterior wall is mainly in white and dark gray. Combination of the white body cut geometrically and unique exterior fin wall is memorable. High-rise SOHO apartment buildings surrounds the shopping mall. Its façade is themed with ocean culture and adopts corrugated exterior projecting slabs to forge the façade shape with strong identity. Meanwhile, as the third generation urban mixed-use of Wanda Plaza, it integrates the one-stop shopping experience consisting of shopping, entertainment and F&B with the program of fashion, leisure and entertainment to provide new shopping experience for people in Xiamen. Xiamen Huli Wanda Plaza started construction in April, 2010 and opened in September, 2011.

广场主立面细部夜景

步行商业街，继续延续了玻璃的底层界面，但高度降低尺度改变，在不断向内延伸中创造出新的城市层次。此种围合，体现由畅到聚，可以作为观赏、群聚、休闲的综合活动场所，从而平衡开放空间利用率并减小局部人流密度

主立面上深咖色铝板于凹槽中镶有灯箱，夜晚灯光效果如深沉天幕上滑落的闪亮陨石群，明暗色彩不断变幻。由电脑操纵的 LED 灯阵产生出无穷的变化，形成种种丰富的图案，巨大的屏幕成为城市中的醒目标志，吸引四方人流汇聚

广场主立面

广场绿化　　　　　　　　广场景观

建筑外立面日景：主立面以深褐色为主色调，沉稳内敛，并有效提升了建筑品质，给人良好的心理感受；但在主要立面上又加入了大面积的浅色铝板，形成强烈的对比，增加视觉冲击，使建筑造型更加醒目。同时，深色建筑体量里穿插着一个个晶莹剔透的玻璃盒子，使整个立面处处透着活泼与灵动。建筑组团注重整体的空间关系及其与城市环境的相融相生的关系，以及与城市天际线的协调；组团内部体块高低错落，疏密有序，空间开合有度。

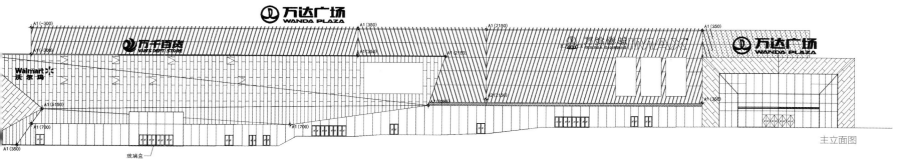

主立面图

建筑外立面局部

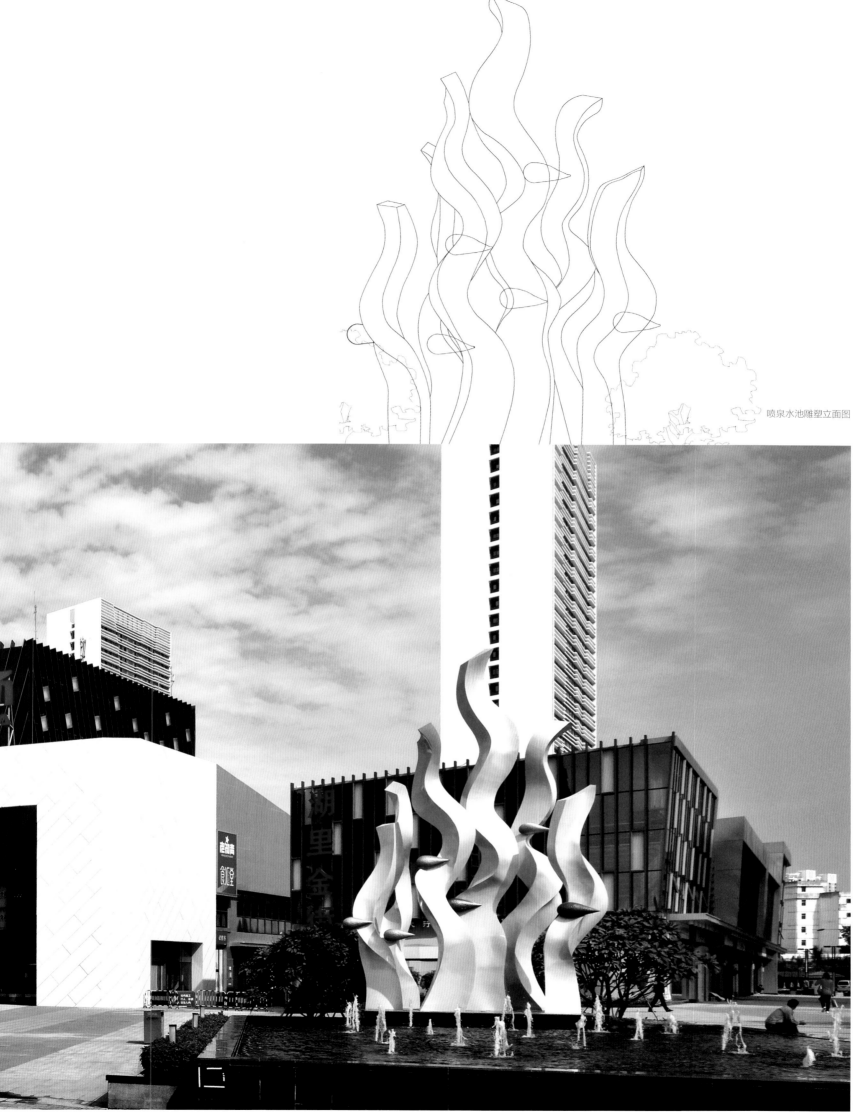

喷泉水池雕塑立面图

主入口广场

室内步行街中庭

室内步行街

室内步行街

大歌星 KTV 平面图

大玩家超乐场

大歌星 KTV 包房

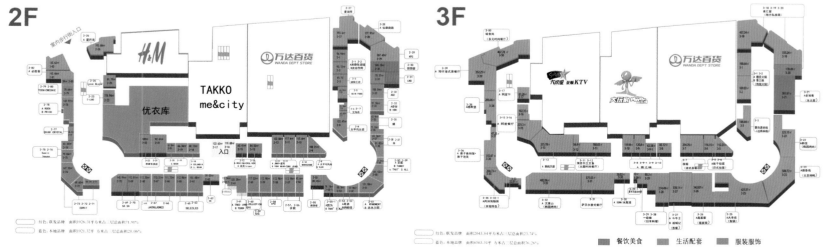

品牌落位图

万达影城大厅

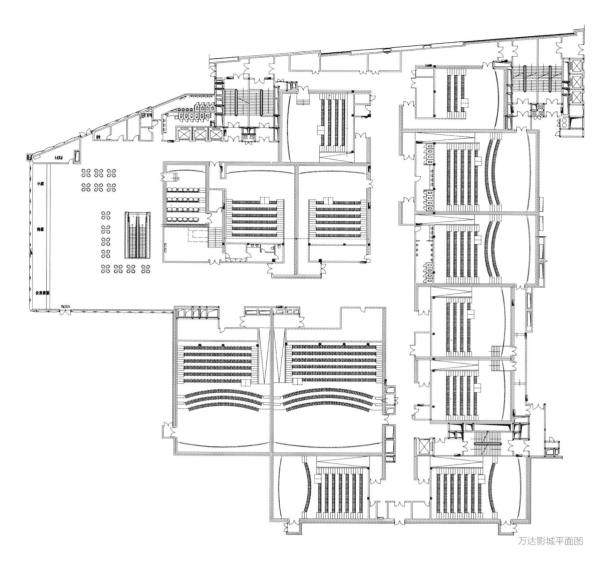
万达影城平面图

万达影城放映厅

万达影城

泰州
万达广场
TAIZHOU WANDA PLAZA

泰州万达广场位于江苏省泰州市凤城河风景区，济川东路以北、海陵南路以西、青年南路以东。总规划用地面积 11.5 公顷，总建筑面积约 41 万平方米，于 2011 年 12 月 9 日盛大开业。项目包含 17 万平方米购物中心、4 万平方米五星级万达希尔顿逸林酒店（250 间客房）、3 万平方米的甲级写字楼、5 万平方米商铺、12 万平方米的精装住宅。该项目的特色为依托于古运河岸而建的酒吧一条街，是市民休闲娱乐的好去处。

大商业东南角

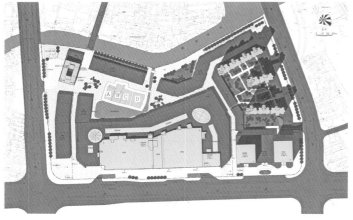

总平面图

The Taizhou Wanda Plaza is located in Fengcheng River side of Taizhou City in Jiangsu Province, south of Jichuan Dong Road, east of Hailing Nan Road and west of the Qingnian Nan Road. The site area is 115,000 m². The gross building area is about 410,000 m². The project opened in Dec 9, 2011.Taizhou Wanda Plaza includes a shopping mall(170,000 m²), Double Trees Hotel by Hilton (40,000 m² and 250 guest rooms), grade A office building (30,000 m²), retail street (50,000 m²) and high quality apartment buildings (120,000 m²). The bar street faces to the historic canal, creating an vibrant and exciting entertainment destination.

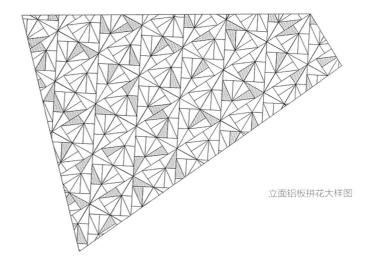

立面铝板拼花大样图

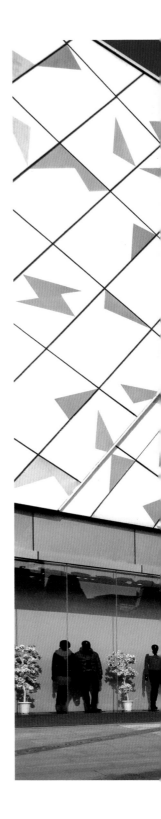

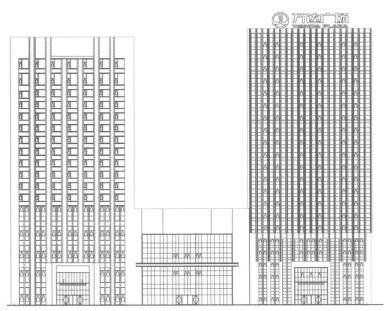

广场局部立面图

广场外立面：外立面不规则几何形体的穿插与对比，强化了商业建筑内在张力与趣味；大体块的虚实对比增强了建筑的独特性

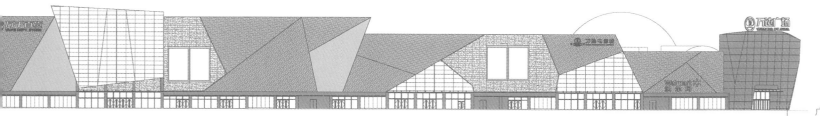

广场立面图

建筑外立面大体块的虚实对比,增强了建筑的独特性

建筑外立面大体块的虚实对比,增强了建筑的独特性

广场外立面

剖面图

大歌星 KTV 平面图

室内步行街中庭

幽雅舒适的购物环境让人流连忘返

室内步行街的吊顶与地面铺装呼应

室内步行街扶梯：运用灯带，营造出抽象的几何空间

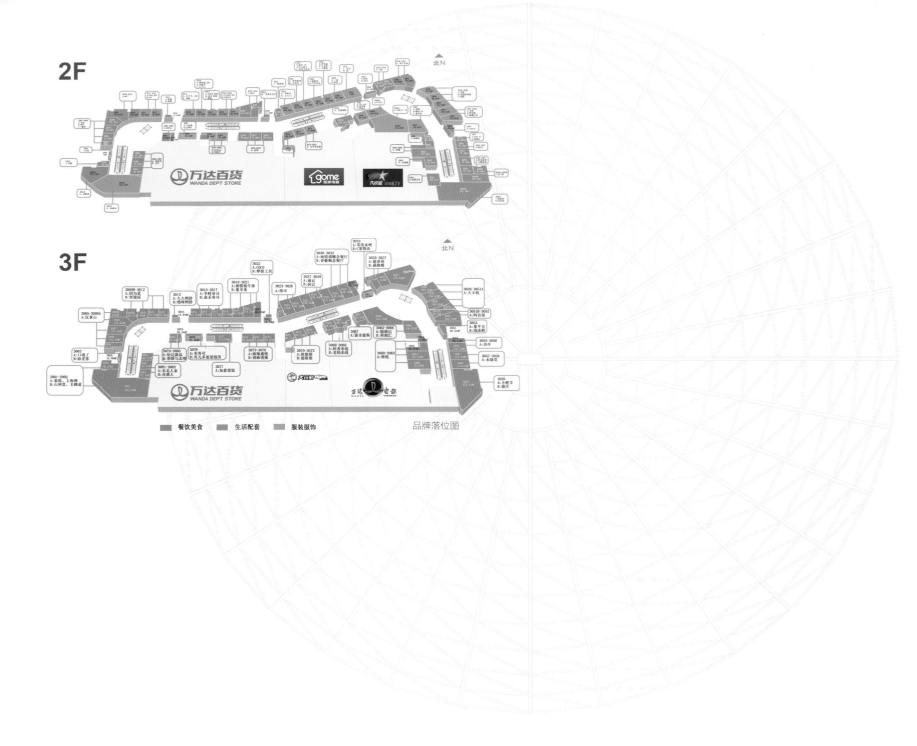

购物中心由室内步行街、百货楼、娱乐楼、大型超市以及与其配套的地下停车场、地下设备用房、管理用房等共同构成，含有万达百货、万达影城、大歌星KTV、大玩家超乐场、国美电器、超市等多种业态。

The shopping center has an indoor pedesctrian street, a department store, entertainment arees, supermarket, as well as other supporting functions, such as underground parking garge, mechanical and administration rooms. The entertainment facilities include Wanda Cinemas, Super Star KTV, Super Player Park, Guomei electronic store, supermarket and other programs.

泰州万达广场的建成开业，极大地提升了泰州的城市品质，增强了区域的商业氛围，并为周边市民提供了便捷的一站式商业服务，带来全新的都市生活体验。

The opening of the "Taizhou Wanda Plaza", will be greatly enhanced the quality of city environment. hugely benefit the regional business climate, and provide a convenient one-stop shopping experience for the surrounding communities and bring a totally new urban experience.

广场主入口夜景

广场沿河夜景

繁华的商业夜景

室内步行街中庭：横向与竖向构图的鲜明对比，自然成为了空间的焦点

唐山路南万达广场
TANGSHAN LUNAN WANDA PLAZA

唐山路南万达广场是万达集团集萃万达23年50余城运营之智，耗巨资为腾飞中的凤凰城带来的真正的国际化、现代化的大型城市综合体，是在新华道的繁华线上描绘的都市不夜城。作为超过110万平方米的超大型城市综合体，集合公寓、商务写字楼、住宅、精品商业街、高档百货、超五星级洲际酒店等6大产品线于一身；荟萃万达百货、大歌星KTV、万达影城、大玩家超乐场等众多商业娱乐业态，打造最具繁华感与国际化的唐山新中心。

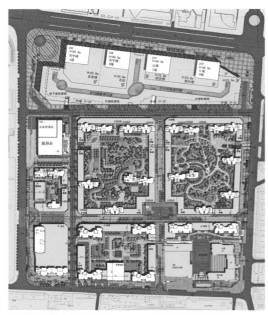

总平面图

广场夜景图

广场西入口门头

购物中心北立面玻璃幕墙背衬满屏 LED 灯具，可展示多姿多彩视频效果，营造强烈商业气氛

Tangshan Lunan Wanda Plaza is an international and modern urban mixed-use projat in the city of Phoenix and urban sleepless city portrayed in Xinhua Avenue, which comes from Wanda Group's intelligence and experience in more than 50 cities for 23 years under huge investment. This ultra-large urban mixed-use project (1.1 million m^2) integrates apartment buildings, office building, residence, boutique business street and upscale merchandises and gathers numerous retail and entertainment programs including super five-star intercontinental hotel, Wanda Dept. Store, Super Star KTV, Wanda Cinemas and Super Player Park to create the most prosperous and international new center of Tangshan.

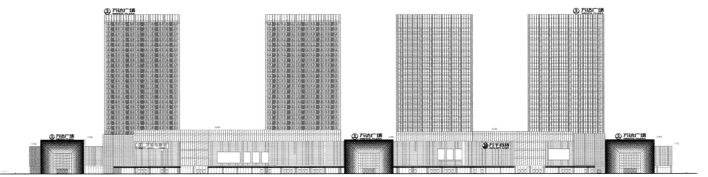

广场主立面图

购物中心门头采用玻璃与铝板虚实变化的设计手法，结合背后 led 灯具，使主入口无论在白天还是夜晚都能牵动眼球，引人入胜。

室内步行街内装

室内步行街中庭　　　　　　　　　　　　　　　　　　　　　　　　室内步行街扶梯

室内步行街中庭活动现场

唐山路南万达广场地处唐山市的核心地带，位于路南区新华道南侧，抗震纪念碑广场东侧，东至增盛路，南临国防道，西至文化路，北至新华道，占地面积 21.27 万平方米，总建筑面积 107.53 万平方米。唐山路南万达广场于 2009 年 5 月 10 日开工，于 2011 年 12 月 23 日开业。

Tangshan Lunan Wanda Plaza is located in the core area of Tangshan City, to the south of Xinhua Avenue of Lunan District and east of Anti-seismic Monumental Square. It occupies 212,700 m² land with gross floor area of 1,075,300 m², bordered by Zengsheng Road to the east, Guofang Avenue to the south, Wenhua Road to the west and Xinhua Avenue to the north. Tangshan Lunan Wanda Plaza started construction in May 10, 2009 and opened in December 23, 2011.

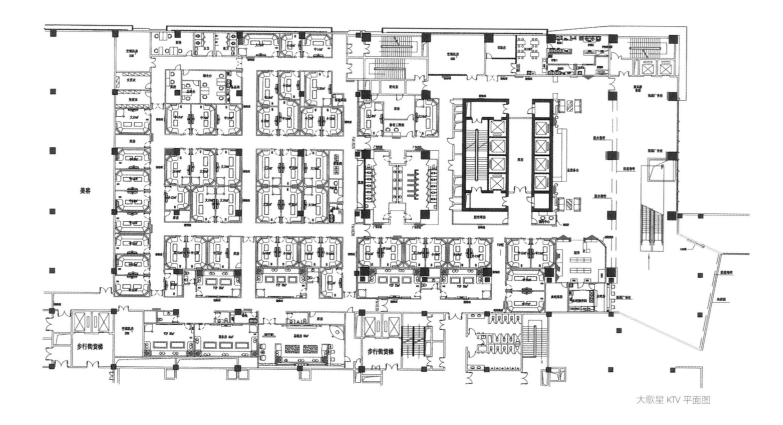

大歌星 KTV 平面图

广场活动现场

广场活动现场

1F

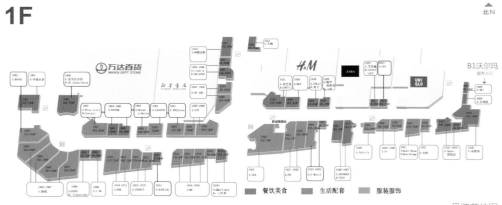

品牌落位图

常州新北万达广场
CHANGZHOU XINBEI WANDA PLAZA

常州新北万达广场位于常州市新北区，用地位于通江大道西侧，北至巢湖路，南至太湖路，西至竹山路。规划用地面积 7.44 公顷，总建筑面积 38.83 万平方米，其中地上建筑面积 29.8 万平方米。容积率 4.0，建筑密度 60%，建筑高度 100 米。功能包括购物中心、五星级酒店、室外商业街、公寓、写字楼及商务酒店。

Changzhou Xinbei Wanda Plaza, with total floor area of 388,300 m², is located in Xinbei District of Changzhou City. The site of the project is about 7.44 hectares, defined by the Riverside Avenue to the west, Nest Lake Road to the north, Tai Lake Road to the south and the Banboo Hill Road to the west. The FAR ratio is 4.0, with building coverage of 60% and above ground floor area of 298,000 m². This business and commercial complex accommodates shopping center, five-star hotel, commercial street, high-end office buildings and a business hotel.

外立面设计以水为主题——购物中心，写意"水波粼粼"；五星级酒店，写意水"倾流而下"。

The facade design of the shopping center and the five-star hotel follows the concept of cascading and waving reflection of waterfall to achieve pleasant and lively commercial effect.

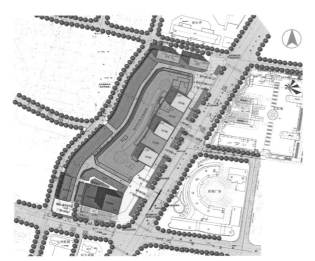

总平面图

夜景效果图

室内步行街中庭采光顶

室内外街联通道

室内步行街连桥

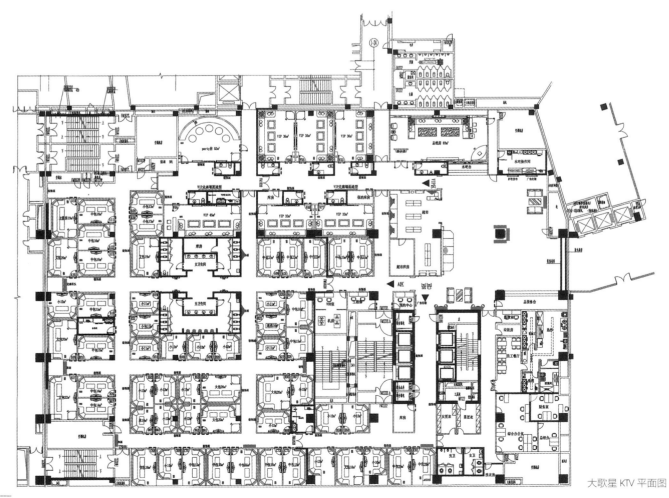
大歌星 KTV 平面图

室内步行街

1F

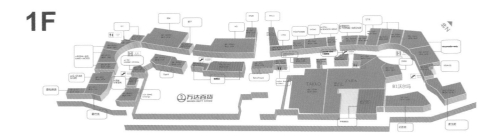

2F

■ 餐饮美食　■ 生活配套　■ 服装服饰

品牌落位图

夜景照明，近5000平方米的LED屏，20余段视频，包括江南意象、水墨风荷、国庆、圣诞等等，充分体现江南地域文化和商业特色。

There is a 5000 m² LED screen with more than 20 videos projecting images of Southern China's landscape, lotus, Festivals of Christmas, National Holidays to represent the local culture of southern China and create the commercial atmosphere.

常州新北万达广场于2010年7月20日开工，于2011年12月10日开业。

The project started to construct on July 20, 2010 and opened on December 10, 2011.

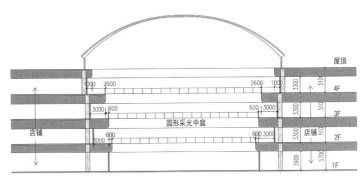

圆形采光中庭

灯具隐藏在立面肌理中，形成近5000m²的视频墙面，变幻着包括"水墨风荷""江南意向"在内的20段视频，呈现了常州的地域特色、文化传承，营造了欢乐、吉庆的商业氛围，具有强烈的振撼力

室外步行街8种标准模块的穿插组合，街巷蜿蜒形成江南小街，步移景换的意向

室内步行街中庭

074　万达广场　WANDA PLAZAS　室内步行街扶梯

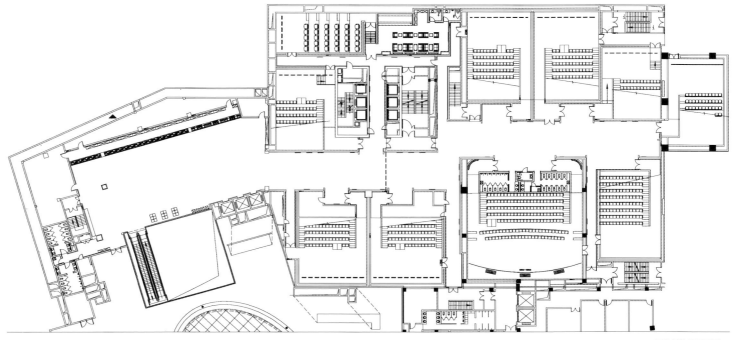

万达影城平面图

室内步行街入口　　　　　　　　　　　　　　　　室内步行街入口

银川金凤万达广场
YINCHUAN JINFENG WANDA PLAZA

银川金凤万达广场位于银川市金凤区正源北街西侧，上海路以南，育安巷以北，正源西街以西。总规划用地面积 7.01 万平方米，其中建设用地 5.67 万平方米，其他代征绿化用地 1.3 万平方米。总建筑面积约 28.5 万平方米，其中地上部分 28 层 21.5 万平方米，地下部分 2 层 7.0 万平方米，建筑高度 100 米。

Yinchuan Jinfeng Wanda Plaza is situated in Jinfeng District, Yinchuan. Surrounded by North Zhengyuan Street on west, Shanghai Road on south, Yu'an Alley on north and West Zhengyuan Street on west, the project covers total area of 70,100 m^2, of which 56,700 m^2 is building area and 13,000 m^2 is public green coverage. The gross floor area is 285,000 m^2, of which there are 28 floors above grade which occupy 215,000 m^2 and 2 floors below grade which occupy 70,000 m^2. The total building height is 100 meters.

广场入口夜景

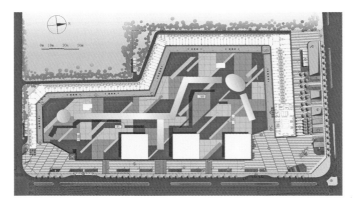

总平面图

夜晚的银川金凤万达广场给这座北方城市增添了一抹亮色,主入口星星点点的光亮,吸引着市民进入这座休闲娱乐购物天堂

广场入口

广场外立面整体夜景

广场活动现场

广场外立面

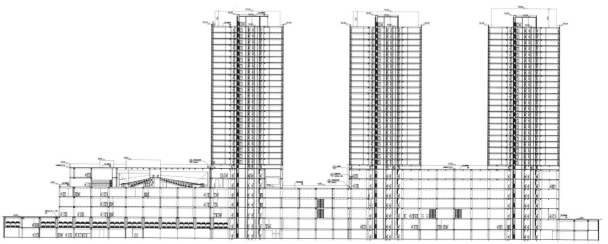

广场剖面图

项目于 2011 年 9 月 9 日盛大开业，购物中心由室内步行街、百货楼、娱乐楼、大型超市以及与其配套的地下停车场、地下设备用房、管理用房等共同构成，含有万达百货、万达影城、大歌星 KTV、大玩家超乐场、国美电器、超市等多种业态。

The plaza opened on September 9, 2011. The shopping center includes an indoor pedestrian shopping street, a department store, entertainment areas, a supermarket and underground garage, mechanical and administration rooms. The project consists of Wanda Dept. Store, Wanda Cinemas, Super Star KTV, Super Player Park, Gome, supermarkets and other programs.

室内步行街

广场丰富的雕塑作品给整个硬朗的建筑增添了活力，市民或驻足观赏，或游戏拍照

室内步行街椭圆中庭

室内步行街内景

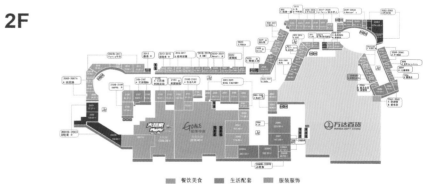
品牌落位图

室内步行街内景

室内步行街内景

室内步行街内景

镇江
万达广场
ZHENJIANG WANDA PLAZA

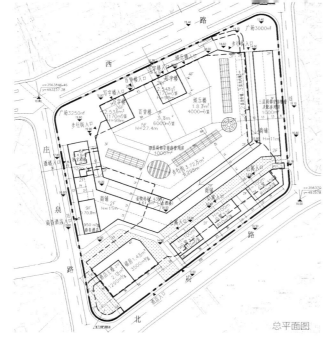

总平面图

镇江万达广场位于江苏省镇江市润州区核心板块,项目建筑面积约39万平方米(地上约30万平方米)。

Zhenjiang Wanda Plaza, as a city complex developed by Wanda Group, is located in the central area of Runzhou district of Zhengjiang City, Jiangsu Province. The grass floor aren is 390,00 m², of which above-grade floors occupy 300,00 m².

镇江万达广场距沪宁城际镇江站和长途客运中心仅一步之遥,是以购物中心、五星级酒店、公寓、室外步行街等多种物业类型为主,构成大体量、"一站式"、多业态的商业格局,是集购物、娱乐、餐饮、办公、居住为一体的综合性开发项目。

Within only blocks away from Shanghai-Ningbo high speed train station and bus terminal, Zhengjiang Wanda Plaza opens integrates building complex of commercial and residential facilities, such as shopping center, 5 star hotel, apartments buildings, shopping street and office buildings.

广场外立面日景

广场入口夜景

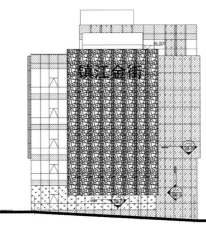

金融街广场方向立面图

室内步行街中庭

室内步行街扶梯　　室内步行街中庭，几何纹样在步行街中延伸

定位为"镇江新商业中心"的镇江万达广场，给镇江带来一种时尚的小资生活。随着万达百货、室内步行街、五星级喜来登酒店的开业，一些从未入驻镇江的高端及中高端品牌也随之亮相，ZARA旗舰店、莎莎、万宁等把镇江的时尚度提升到一个新的水平。

Zhenjiang Wanda Plaza, valued as "Zhengjiang New Commercial Center", has incorporated the fashionable petty life into Zhengjiang people's life. With the opening of Wanda Dept. Store, shopping street, and the Sheraton Hotel, high end brands such as Zara flagship, Sasa, Mannings, also set up their shops in Zhenjiang, and provide a fashionable shopping environment for the city of zhenjiang.

规模宏大的体量、简洁现代的外立面、多彩炫目的夜景照明给城市带来新的气象，明亮时尚的室内步行街营造出良好的购物环境，丰富多彩的业态给大众带来的全方位的休闲体验。镇江万达广场从开业起就一直成为人潮涌动的欢乐地。

Large building scale, modern façade with colorful lighting illumination and variety of services has greatly enhanced the city's shopping and entertainment experiences. From the opening day, Zhengjiang Wanda Plaza has been the hub for people, business and commerce in this city.

镇江万达广场于2009年12月29日开工，于2011年8月12日开业。

Zhenjiang Wanda Plaza started construction on December 29, 2009 and opened on August 12, 2011.

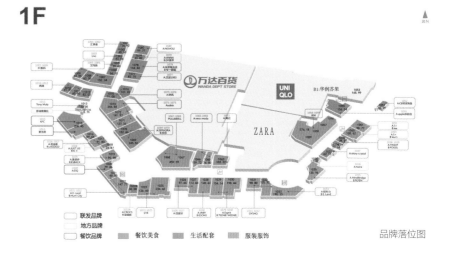

品牌落位图

武汉经开万达广场
WUHAN JINGKAI WANDA PLAZA

武汉经开万达广场用地位于武汉沌口经济技术开发区 318 国道东侧，体育路西侧，南与开发区体育中心隔路相邻，项目总规划用地面积约 12 万平方米，总建筑面积约 44.12 万平方米。

Wuhan Jingkai Wanda Plaza is located in Wuhan Dunkou Economic and Technological Development Zone, east to the No.318 State Road, west to the Sports Road, and south to the Sports Center which is at the other side of the road. The overall project planning land area is about 120,000 m², and the total construction area is about 441,200 m².

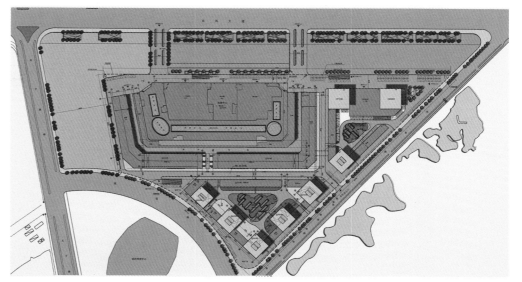

总平面图

广场外立面

项目针对经济技术开发区城市功能总体要求及市场需求，规划有约 16 万平方米的购物中心，万达百货、万达影城、国美电器、大型超市、大歌星 KTV 等诸多主力店尽数落户于此，是经开区首家一站式全程购物的休闲场所；商务区包括五星级酒店、写字楼、商务酒店等功能，其中五星级酒店建成后将成为经开区第一家集会议、接待、餐饮为一体的豪华酒店，必将大大提升万达广场乃至经开区的整体商务形象。

For the purpose to meet the overall demands of the Economic and Technological Development Zone, the project plans to have total GFA about 160,000 m², which includes shopping center, Wanda Dept. Store, Wanda Cinemas, GOME, Wal-Mart, Super Star KTV and other. It is the first one-stop shopping and leisure complex. Its main business district includes 5-star hotels, business hotels, office building. Among these facilities, the five-stars hotel, in the Jingkai Wanda will be the first luxury hotel to host conference, reception, and catering events. It will greatly enhance the Wanda Plaza and Jingkai's overall business imagery.

武汉经开万达广场的入驻，在填补了武汉经济开发区产业升级所带来的商业发展空白的同时，更为当地百姓提供了巨大的生活便利和投资机遇，翻开了武汉市商业发展的又一崭新篇章。

The opening of Wuhan Jingkai Wanda Plaza will not only upgrade the overall business environment in Wuhan Economic Development Zone, further more, it provides a great convenience and investment opportunities for the local people, which create the new chapten of the business development in Wuhan.

广场夜景图——五光十色的商业照明,寓示了商业的繁华与兴旺

室内步行街中庭：空间的音符与旋律让人联想到建筑是凝固的音乐

1F

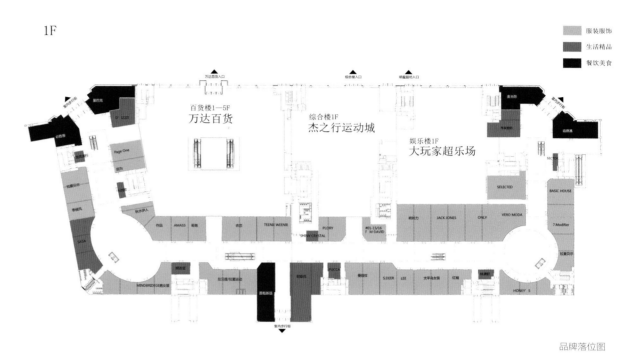

品牌落位图

建筑的韵律美让人心旷神怡

室内步行街中庭　　　　　　　　室内步行街扶梯

上海江桥
万达广场
SHANGHAI JIANGQIAO
WANDA PLAZA

广场外立面夜景

广场外立面

总平面图

广场外立面夜景

上海江桥万达广场位于上海市嘉定区江桥镇,基地南至鹤旋路,东至华江路,西至黄家花园路,北至金沙江西路。

Shanghai Jiangqiao Wanda Plaza is located in Jiading District of Jiangqiao Town in Shanghai, south to the Hexuan Road, east to the Huajiang Road, west to Huangjia Huayuan Road, and north to the Jinsha Jiangxi Road.

江桥万达广场由百货楼、综合楼、超市楼、娱乐楼和室内步行街、室外步行街及办公楼等组成,包括超市、百货、电玩、影院、电器、KTV等各大主力店。其中,商业裙房高度为2~5层,11栋塔楼高度为10~12层,建筑总高度不超过48米。地下室为1层。上海江桥万达广场于2009年11月21日开工,2011年6月18日开业。

This project is a large comprehensive commercial complex which includes a department store, a muli-function building, a supermarket, an entertainment building and an indoor and outdoor pedestrian street, and office buildings. It also includes supermarkets, department stores, super player Park, cinemas, appliances stores, KTV, and other major anchor stores. The retail plinth is about two to five stories high. The 11 tower buildings are about ten to twelve stories high. The building's total height is 48 m. The basement has one story. Shanghai Jiangqiao Wanda Plaza started construction on November 21, 2009 and opened on June 18, 2011.

广场活动现场

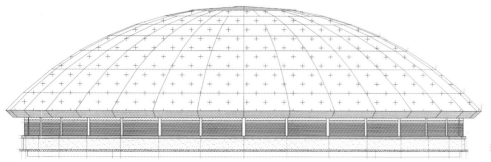

采光顶大样

室内步行街中庭：运用形式各异的采光顶篷，塑造宽敞、明亮的室内商业街形象

室内步行街中庭

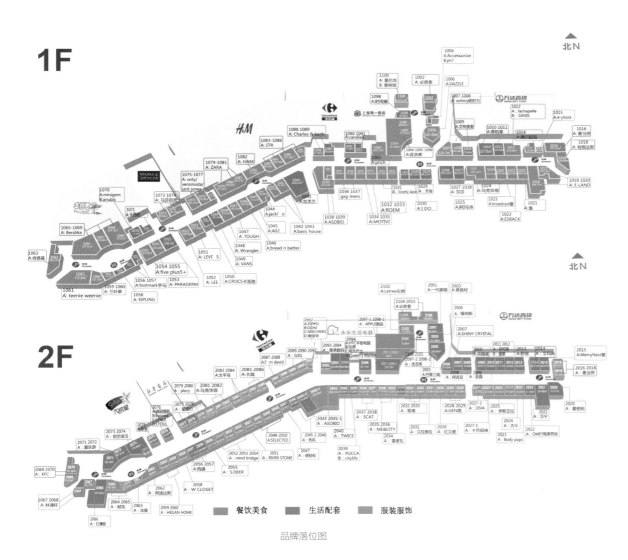

品牌落位图

室内步行街中庭　　　　　　　　　　　　　　　　　　　　　　　　室内步行街

室内步行街扶梯

廊坊
万达广场
LANGFANG WANDA PLAZA

廊坊万达广场位于河北省廊坊市，东至新华路，西至文明路，南至金光道，北至永丰道。地块总用地面积约 13.5 万平方米，由商业综合体、五星级酒店、5A 写字楼、室外商业街、商务酒店、住宅及底商组成。总建筑面积约 69.5 万平方米，其中地上建筑面积约 53.8 万平方米，地下建筑面积约 15.7 万平方米。

Langfang Wanda Plaza is located in the core area of Hebei Langfang City. The site connects Xinhua Road on east, Wenming Road on west, Jinguang Road on south and Yongfeng Road on north. The site area is about 135,000 m², including large scale commercial complex, five-star hotel, Grade A office building, outdoor pedestrian street, business hotel, and other retail programs. This project integrates shopping, entertainment, leisurement, cuture, housing and so on into a large-scale urban complex. The gross floor area is around 695,000 m², including 538,000 m² above ground and 157,000 m² in basement.

广场外立面

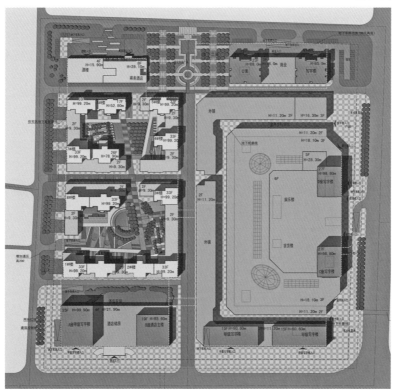

总平面图

广场入口夜景

广场雕塑

广场东北角日景

广场夜景

广场夜景局部

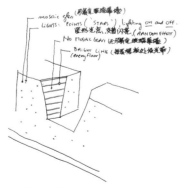

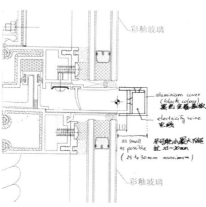

大师草图　　大师草图：彩釉玻璃安装示意图

廊坊万达广场沿城市主干道的"马赛克"玻璃立面，在不同的日照条件下，给人不同的绚丽奇幻的视觉冲击，尤其是在日出、日落之时。这种大面积玻璃的运用和表现手法在世界范围内还没有先例，体现了万达敢于创新的精神。廊坊万达广场于2010年10月开工，2011年11月开业。

The large "mosaics" glass façade facing the main street gives pedestrian shocking and illusive visual impression under different sun orientation, especially during sunrise and sunset. The usage of mosaics glass in such a large area in building façade has never been seen in the world. It indicates the spirit of innovation of Wanda Group. Langfang Wanda Plaza started construction in Octerber. 2010 and opened in November, 2011.

广场西南主入口

室内步行街入口

室内步行街主入口

室内步行街连桥

品牌落位图

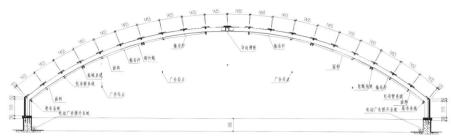

椭圆采光顶剖面图

室内步行街内装

室内步行街大中庭　　　　　　室内步行街扶梯

郑州中原万达广场
ZHENGZHOU ZHONGYUAN WANDA PLAZA

郑州中原万达广场位于河南省郑州市中原区，华山路东、秦岭路西、中原西路南、伊河路北约100米所围合成的地块内，总用地面积为9.27万平方米。

Zhengzhou Zhongyuan Wanda Plaza is located in the Zhongyuan District of Zhengzhou City in Henan Province, east to Huashan Road, west to Qinling Road, and north to Yihe Road with in 100 m. Its total site area is 92,700 m².

项目规划总建筑面积约53.1万平方米，其中地上42.1万平方米，地下11万平方米，由购物中心、写字楼、室外步行街、外铺、住宅及底商组成。

The total planning GFA is about 531,000 m², of which the above-ground floor area is 421,000 m², and the underground floor area is 110,000 m², including shopping centers, office buildings, outdoor pedestrian street, along with shopping stores and residential buildings.

广场外立面夜景

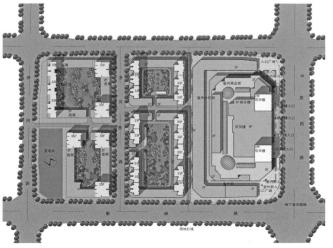

总平面图

广场外立面

室内步行街大中庭

品牌落位图

以城市总体规划为依据,通过研究城市设计、商务、办公、商业、居住及休闲活动等空间模式和景观环境各要素,从现代城市功能、景观要求出发,对规划用地进行整体设计,塑造"城市中心地标建筑"。郑州中原万达广场于2010年7月10日开工,于2011年10月28日开业。

The overall master plan design is based upon city planning guideline under the comprehensive studies of city planning, business, and office environment, commercial, residential and leisure activities, spatial pattern and landscape elements, and it will be shaped as "the landmarketed building of the city center". Zhengzhou Zhongyuan Wanda Plaza Project started construction on July 10, 2010 and opened on October 28, 2011.

室内步行街　　　　　　　　　　　室内步行街

室内步行街

石家庄裕华万达广场
SHIJIAZHUANG YUHUA WANDA PLAZA

石家庄裕华万达广场位于石家庄市裕华区，南至槐安路，北至槐中路，西至民心河西侧，东至建华大街西侧。规划总建筑面积约183.90万平方米，是集大型购物中心、五星级洲际酒店、娱乐中心、城市商业步行街、国际标准5A写字楼、精装公寓、高端豪宅为一体的首席城市综合体。

Shijiazhuang Yuhua Wanda Plaza is situated in Yuhua District, south to Huaian Road, north to Huaizhong Road, west to Minxin River and east to Jianhua Avenue. The gross floor area is 1.83 million m². This mixed-use urban complex includes shopping center, five-star Intercontinental Hotel, entertainment center, commercial pedestrian street, 5A office building, fine-deco apartment and luxury residence.

远眺街区，使人充分感受商业"巨无霸"的恢宏气势

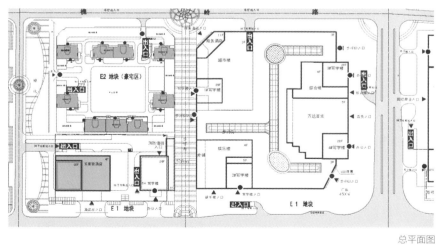

总平面图

广场东南角夜景

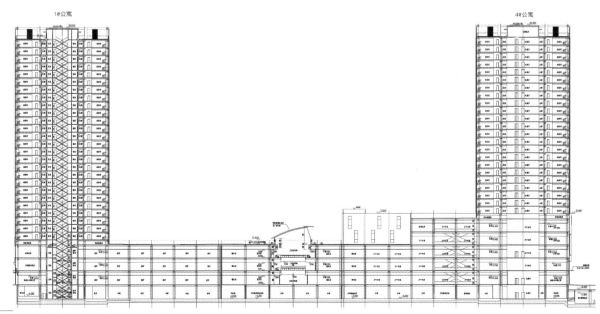

剖面图

商业入口夜景：尺度夸张的入口彰显现代商业建筑的时代氛围

五彩的琴键饰以精致的金属幕墙，体现"建筑是凝固的音乐"的最高境界

广场局部外景

石家庄裕华万达广场引进多家国际时尚品牌，如UNIQLO、C&A、H&M、ZARA等，成为名副其实的河北省高端商业综合体。

Shijiazhuang Yuhua Wanda Plaza imported many international fashion-style brands such as Uniqlo, C&A, H&M and ZARA etc. It becomes the top-notch mixed-use complex in Hebei Province.

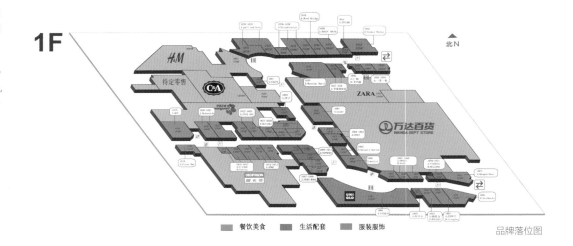

品牌落位图

室内步行街内装

室内步行街扶梯

二层长街

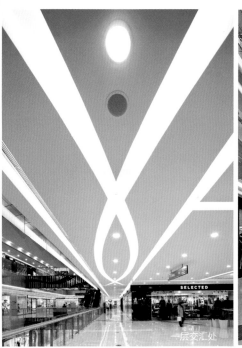

一层交汇处

室内步行街

室内步行街中庭

大庆萨尔图
万达广场
DAQING SAERTU WANDA PLAZA

大庆萨尔图万达广场是目前大庆唯一的城市综合体项目，是万达集团在大庆布局的第一个万达广场，同时也是目前中国最北端的万达广场，黑龙江省最大的购物中心。

Daqing Saertu Wanda Plaza is the only mixed-use complex project in Daqing City, and it is Wanda Group's first Wanda Plaza in Daqing as well. Furthermore, it is the most northern Wanda Plaza in China, and the largest shopping center in Heilongjiang Province.

项目总建筑面积 58 万平方米，购物中心总面积 16 万平方米，总投资 30 亿元。项目南邻世纪大道，东邻经三街，西为经二街，北为规划路。城市政治与商业核心区域核心地段，区域交通便利。

The project's total GFA is 580,000 m², including a shopping center of 160,000 m², the total investment is 3 billion RMB. The project is located south to Century Avenue, east to Jingsan Street, west to the Jing'er Street and north to the Guihua Road. It is within city's political and commercial core area with convenient regional transportation access.

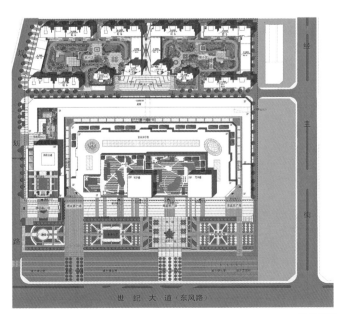

总平面图

广场外立面夜景

广场外立面　　室内步行街中庭：回形中庭空间的光影变化

大庆萨尔图万达广场于 2011 年 11 月 25 日盛大开业。商业广场汇集购物中心、五星级酒店、室内外商业步行街、商务写字楼、高档住宅、精装 SOHO 等多种业态，形成购物、休闲、餐饮、文化、娱乐等多种功能于一体的购物中心。

Daqing Saertu Wanda Plaza grandly opened on November 25, 2011. This commereial complex combines shopping, leisure catering, culture, and entertainment functions together including Shopping Centers, five-star hotels, indoor and outdoor retail pedestrian streets, office buildings, luxury residential towers, and SOHO apartment buildings.

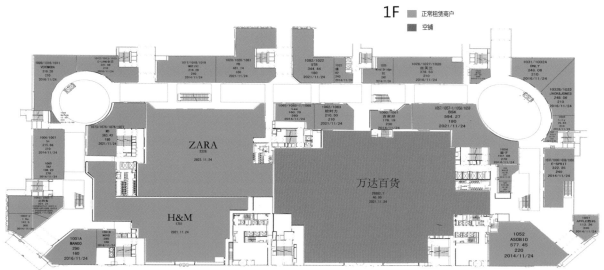

品牌落位图

商业步行街入口

室内步行街扶梯

淮安
万达广场
HUAI'AN WANDA PLAZA

淮安万达广场位于淮安市清河区水渡口地区，分为1号地和4号地两个地块。其中1号地位于水渡口广场东北侧，用地面积约13万平方米，总建筑面积约60万平方米，由购物中心、五星级酒店、甲级写字楼组成；地上37层，地下2层。4号地用地面积约4.8万平方米，总建筑面积为14万平方米，包括10万平方米住宅，1.6万平方米底商，2.4万平方米地下车库。

Huai'an Wanda Plaza is located in the Shuidukou District of Qinghe area in Huaian. It is divided into two separate parcels of 1# and 4#. Parcel 1# locats in the northeast side of Shuidukou plaza, and the site area is about 130,000 m², with a total GFA of 600,000 m², from shopping center, Five-star Hotel, Grade A office building, on the ground 37 layers, underground 2 layers. 4# land area is about 48,000 m², with a total construction area of 140,000 m², including 100,000 m² of housing, 16,000 m² first floor business, 24,000 m² of underground garage.

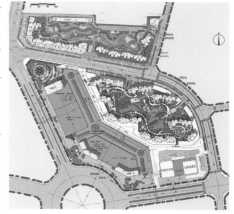

总平面图

广场外立面

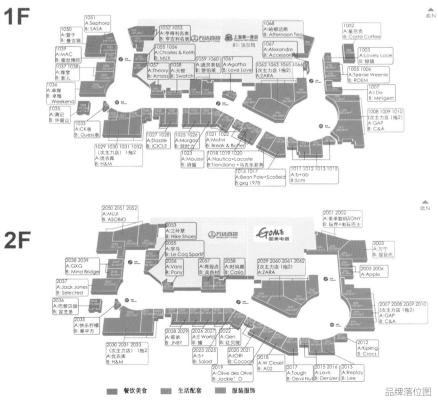

品牌落位图

淮安万达广场于 2011 年 1 月 7 日盛大开业，购物中心由室内步行街、百货楼、娱乐楼、大型超市以及配套的地下停车场、地下设备用房、管理用房等共同构成，包括万达百货、万达影城、大歌星 KTV、大玩家超乐场、国美电器、超市等多种业态。

Huai'an Wanda Plaza opened on January 7, 2011. The shopping center consists of indoor pedestrian street, department stores, entertainment buildings and supermarkets, related underground parking garage, underground equipment rooms, management offices. It includes Wanda Dept. Store, Wanda Cinemas, Super Star KTV, Super Player Park, GOME, supermarket and other programs.

明快、简洁、优雅的商业空间

室内步行街中庭　　　　　　　　　　　　万达影城放映厅

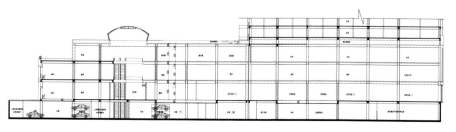
步行街剖面图

PART 2 酒店
HOTELS

万达集团目前已开业 28 家五星和超五星级酒店。万达集团计划到 2015 年开业 80 家五星和超五星级酒店，营业面积 300 万平方米，成为全球最大的五星级酒店业主。万达拥有五星级酒店品牌——万达嘉华、超五星级酒店品牌——万达文华以及顶级奢华酒店品牌——万达瑞华。

Wanda Group has opened 28 five-star and super five-star hotels so far. The Group also plans to open 80 five-star and super five-star hotels by 2015, with an operation area of 3 million square meters, as the world's largest five-star hotel owner. Wanda Group owns its five-star hotel brand–Wanda Realm, super five-star hotel brand–Wanda Vista and top luxury hotel brand–Wanda Reign.

万达集团旗下酒店分为城市类酒店和度假类酒店两大类。万达城市类酒店通常位于城市中心，一般和万达广场等业态组成万达城市综合体。万达度假类酒店通常位于文化旅游区，一般和旅游小镇、秀场、主题公园等组成万达文化旅游项目。

Wanda Group's hotels are divided into two major categories: the urban hotels and the resort hotels. The urban hotels are usually located in the city center, normally forming a Wanda urban complex together with Wanda Plazas, etc.. The resort hotels are usually located in the cultural tourism zone, normally forming a cultural tourism project together with the tourist town, the show theatre, the theme park, and so on.

武汉万达威斯汀酒店
THE WESTIN WUHAN WUCHANG

武汉万达威斯汀酒店坐落于长江之滨，拥有宽阔的视野，可俯瞰迷人的长江美景。酒店高度 99.09 米，层高 3.8 米，共有 305 间客房，总建筑面积 51144 平方米。

The Westin Wuhan Wuchang is located on the bank of the Yangtze River. Its largc windows have wide view and take full advantage of the location's expansive views of the Yangtze River's enchanting beauty. The hotel building is 99.09 m high, with floor to floor height of 3.8 m, gross floor area of 51,144 m^2 and a total of 305 rooms.

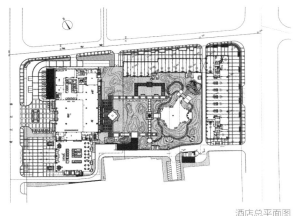

酒店总平面图

气势恢宏的酒店入口

酒店外立面夜景

酒店外立面具有金属质感

酒店立面由上海霍普建筑设计事务所有限公司设计，整体建筑在阳光下呈现一种独特而现代的金属质感和纯净而向上的体量感。

Designed by Shanghai Huopu Architectural Design Consultant INC, the building is a composition of metal and glass, strong steel verticals work in unison with the strong horizontals of the lobby space. The glass dances between the steel verticals, creating a wonderful ever-changing texture, reacting to its surroundings, reflecting the sky and water. The building is most spectacular in full sunlight.

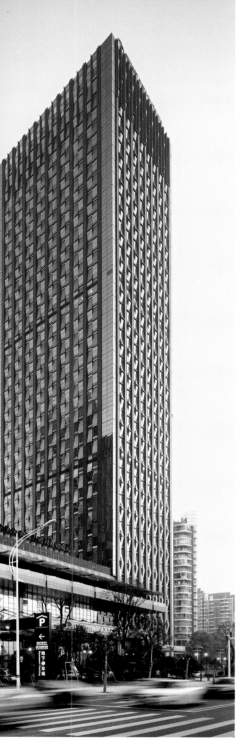

像钻石般具有折线起伏感的建筑立面

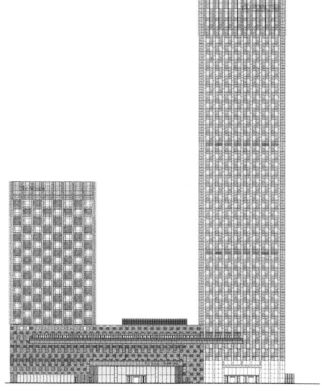

正立面图

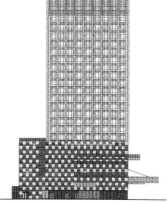

侧立面图

酒店庭院

酒店入口水池

酒店庭院廊道

酒店还拥有面积达12000平方米的欧式后花园，长度达100多米，按照欧洲皇家宫廷园林的设计思想，采用中轴景观的布局，同时结合中国传统园林的"层层递进"的空间关系，给不同区域的使用者以不同的景观感受。武汉万达威斯汀酒店于2009年9月25日开工，于2011年7月30日开业。

The hotel also has a garden in the European style, covering more than 12,000 m², with an elegant proportion reminiscent of the royal palaces of Europe. The landscape is designed along a central axis, including elements of traditional Chinese garden, with its "sequence of spaces" integrating many water features and gazebo, to give users a wide variety of different regions and landscape feelings. The Westin Wuhan Wuchang started construction on September 25, 2009 and was opened on July 30, 2011.

酒店庭院　　　　　　　　　　　　大堂吧室外平台

全日式餐厅：现代材料与传统元素的结合，空间显得精致与空透

雍容华贵的酒店大堂

酒店大堂

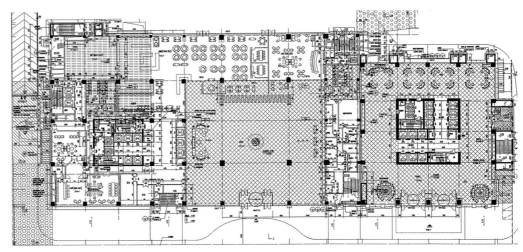
酒店首层平面图

广州白云万达希尔顿酒店
HILTON GUANGZHOU BAIYUN

广州白云万达希尔顿酒店总建筑面积 4.5 万平方米 地上约 3.7 万平方米。酒店地处白云山脚，在静谧舒适的自然环境中创造出纯净精致的现代生活空间，拥有 313 套舒适优雅的客房，6 间风格各异的餐饮设施和酒吧，10 个世界级水准的会议及宴会厅与高品位的 SPA 和健康中心。

Hilton Guangzhou Baiyun has a gross floor area of 45,000 m² with a total floor area above groud of 37,000 m². Located in the foot of Baiyun mountain. It forms a clean and delicate modern living space in the peaceful and comfortable natural environment. It has 313 comfortable and elegant guest rooms, 6 restaurants and bars in different styles, 10 world-class meeting and banquet rooms, as well as a top SPA and fitness center.

酒店入口

酒店日景

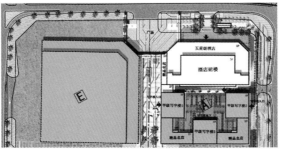

总平面图

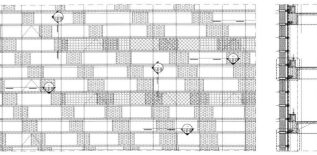

酒店幕墙大样

酒店的会议及宴会场地运用了灵活的设计，扶梯可直达二楼10个多功能厅，包含两个贵宾室和一个私密的新娘化妆间，占据3220平方米，还有一个壮观典雅的8.5米无柱式大宴会厅占据1500平方米。在装潢及质感选材方面使用淡金渐变象牙色大理石及原木，以营造现代舒适的氛围。房间设有落地玻璃窗，令室内光亮通透，让宾客置身繁荣商业区中鸟瞰白云翠景、俯瞰迷人城市景观，并呼吸清新自然的空气。

The design of conference and banquet areas maximizes their flexibility, with an escalator direct to 10 multifunctional halls on the second floor, including two VIP rooms and a private bride dressing room which occupies 3,220 m². It also has a grand and elegant astylar banquet hall with 8.5m floor to floor high and total area of 1,500 m². Logs and Ivory tint marbles with light golden shades are selected as the texture for decoration, to create comfortable modernized atmosphere. French windows make the room bright and transparent, and also allow guests overlook the jade green and charming urban landscape, and breathe fresh air in the flourishing business area.

酒店整体体量完整、外立面设计极具现代感，夜景照明结合立面设计特点炫彩灵动。万达希尔顿酒店于2010年2月11日开工，2011年8月13日开业。

The hotel has an integrated volume and modern exterior design. Facade lighting design utilizes the feature of facade design to create colourful and spiritual effect. Hilton Guangzhou Baiyun started construction on February 11, 2010 and opened on August 13, 2011.

宴会厅豪华包间

酒店游泳池

酒店豪华宴会厅

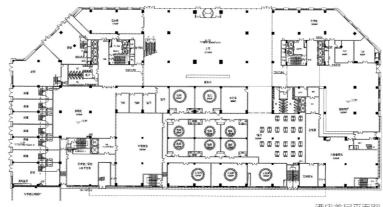

酒店首层平面图

济南万达凯悦酒店
HYATT REGENCY JI'NAN

济南万达凯悦酒店，是济南魏家庄万达广场城市综合体重要的组成部分，位于济南市市中区经四路以北，顺河街以西，经二路以南，纬一路以东。总建筑面积5.39万平方米，其中地上建筑面积4.36万平方米，总客房数345间，酒店大堂800平方米，宴会厅1000平方米，是济南中心区高端酒店的代表。

Hyatt Regency Ji'nan is an important part of Ji'nan Weijiazhuang Wanda Plaza, an urban complex located to the north of Jingsi Road in Shizhong district of Jinan, to the west of Shunhe Street, to the south of Jinger Road and to the east of Weiyi Road. The hotel has a gross floor area of 53,900 m² and a total floor area of 43,600 m² above ground, with 345 guest rooms, a 800 m² hotel lobby as well as a 1000 m² banquet hall. This hotel represents one of the top hotels in the city centre of Ji'nan.

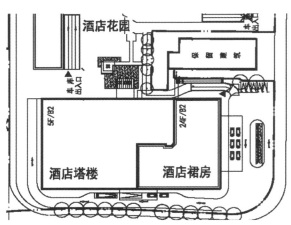

酒店总平面图

大尺度出挑雨篷，精致的细节处理，加之夜景照明设计，带给人奢华的空间感受

酒店外立面夜景：竖向线条、米色石材与玻璃幕墙相结合，尽显酒店的高贵气质

酒店日景

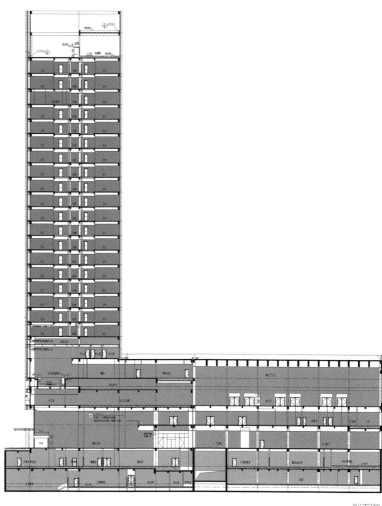

剖面图

外立面通过米色石材与玻璃幕墙的结合，充分体现济南当地北方建筑特色。

Its exterior is designed with an combination of cream-coloured stone and glass curtain wall, which presents the local architecture style in northern area.

室内游泳池

内装设计充分挖掘地方文化元素,彰显典雅、大气的风格。

The interior design makes full use of local cultural elements which is delicate and generous.

景观设计营造静谧、高贵的氛围,并实现了与商业广场的有效分隔与联系。济南万达凯悦酒店于 2009 年 6 月 10 日开工,于 2011 年 11 月 18 日开业。

The landscape design gives a sense of peace and nobility. Meanwhile it makes a successful division and connection with the commercial plaza. Hyatt Regency Ji'nan started constuction on June10. 2009 and opened on November 18. 2011.

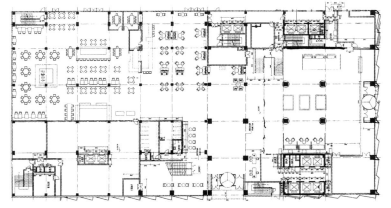

酒店首层平面图

酒店大堂

电梯厅

大宴会厅

中式餐厅

南京万达
希尔顿酒店
HILTON NANJING

秉承万达集团锐意超越的精神，万达广场在高端配备上不断完成历史性跨越。南京万达倾力打造的商务区巨擘之作——南京万达希尔顿酒店，于2011年11月16日正式开业。项目位于南京市建邺区江东中路100号，地理位置优越。

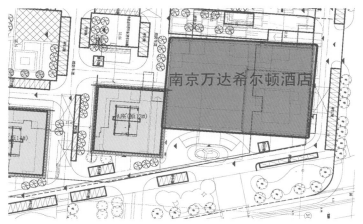

酒店总平面图

餐厅：个性与私密的高级场所

By inheriting the outstanding mission from Wanda Group, Wanda Plaza continuously construct the high-level facilities to create the historical wonderland in Nanjing. Wanda Nanjing spared no effort to build Hilton Nanjing, which can be seen as a brilliant star in the newly developed CBD. Hilton Nanjing, opened on November 16, 2011, conveniently located in No.100 Middle Jiangdong Road, Jianye District.

Art deco 风格的竖向线条让建筑高耸挺拔

酒店立面日景

酒店立面夜景

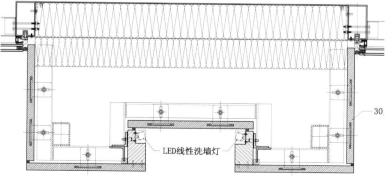

灯具暗藏节点

南京万达希尔顿酒店坐拥 350 间客房,全新理念设计的中西式餐厅及酒吧,宽敞舒适的会议场地及宴会空间,健身设施涵盖泳池和健身房,一应俱全的配套服务,能充分满足客人商务休闲的需求。

Hilton Nanjing is a 350-room hotel hosting innovative restaurants and bars, extensive meeting and banqueting space, and fitness facilities including a pool and health club.

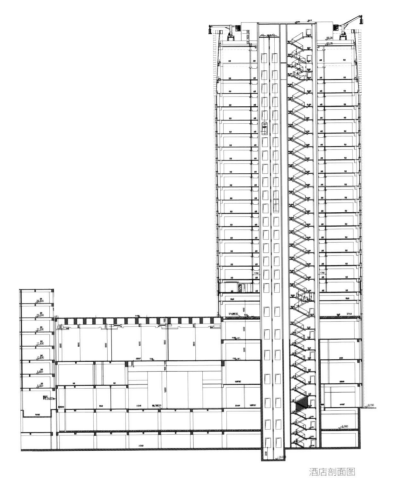

酒店剖面图

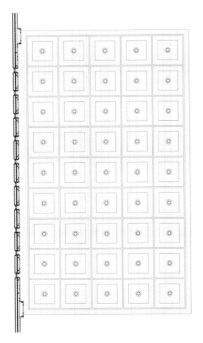

雨篷顶视图

大尺度的建筑入口兼具功能与美观的双重功能

大堂吧中金属马赛克在灯光的照耀下熠熠生辉

1200平方米的无柱式大宴会厅以及七个独立可拆分组合的多功能厅，配有音频/视频设备、会议电话、网络会议设施、商务电话服务、快件以及复印和打印设备。可容纳1200人的大宴会厅作为南京最大的宴会厅之一，适合举办各类会议及宴会。

The hotel owns a 1,200 m² pillar-less grand ballroom with 7 multi-function rooms, with A/V rental centre, conference phone, internet conference facilities, business phone services and express mail as well as photocopying and printing equipment. As one of the biggest grand ballrooms in Nanjing, with a capacity of 1200 people, it is a perfect place to provide catering for all types of conferences and banquets.

酒店大堂：优雅奢华的空间感受

宴会厅

总统套房

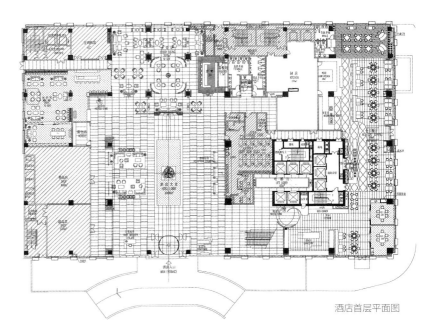

酒店首层平面图

唐山万达洲际酒店
INTERCONTINENTAL TANGSHAN

唐山万达洲际酒店坐落于唐山主要干道新华道与文化路交汇的城市核心地带，坐东朝西，东依路南万达广场，背靠万达居住区，面向广阔的抗震公园，地理位置得天独厚。

Intercontinental Tangshan is located in the core area where the arterial road Xinhua Road and Wenhua Road intersect. It is east-west oriented, east to Wanda Plaza, at the back of Wanda residential zone, facing the broad Anti-seismic Memorial Park, with excellent geographic location.

酒店总建筑面积 4.77 万平方米，其中地上 3.64 万平方米，地下 1.13 万平方米。秉承万达豪华酒店的一贯特色，唐山万达洲际酒店是具有 298 间客房，含部长套、总统套、行政酒廊、室内泳池等现代化居住娱乐设施，荟萃中西餐饮的高档五星级酒店。

The gross floor area of hotel is 47,700 m², where the above-ground area is 36,400 m² and underground area is 11,300 m². Following the traditional features of Wanda luxury hotel, Tangshan Wanda Intercontinental Hotel has 298 guest rooms, including minister suite, president suite, administrative lounge, indoor swimming pool and other modern living and recreational facilities. It is a high-end five-star hotel gathered with different styles of Chinese restaurant and western restaurants.

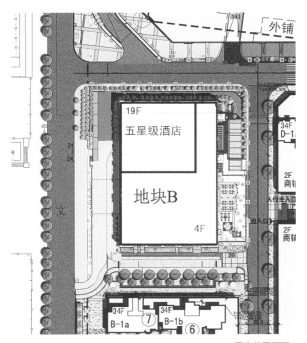

酒店总平面图

酒店外立面

酒店地上19层,立面设计以竖向线条为主,体现万达建筑一贯的简洁、现代、大气的设计风格,展现出企业独特内在文化。唐山万达洲际酒店材质高贵,体量高耸,内装奢华,独具一格,是万达集团为唐山市奉献的具有真正国际品味和管理水平的豪华商务场所。唐山万达洲际酒店于2009年3月开工,于2011年12月23日开业。

The hotel has 19 floors above ground. The vertical linear facade reflects the Wanda design style of succinctness, modern and grandeur unique enterprise culture. Intercontinental Tangshan is a luxury commercial and busiess site of Tangshan city with noble temperament, erecting profile, luxury interior fit-out, unique style and international taste and management standard. Intercontinental Tangshan started construction in March 2009 and opened on December 23, 2011.

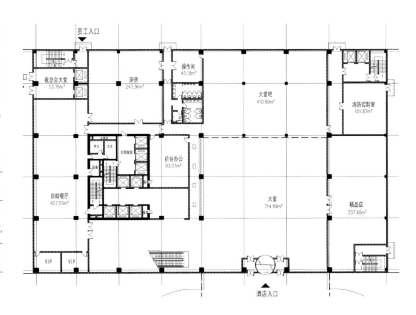

酒店首层平面图

从抗震公园看酒店

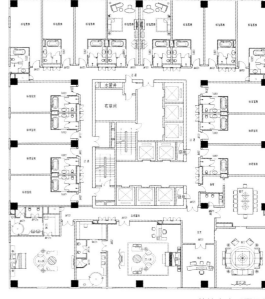

总统套房层平面图

客房

VIP吧

石家庄万达洲际酒店
INTERCONTINENTAL SHIJIAZHUANG

石家庄万达洲际酒店位于河北省石家庄市裕华区，南临槐安路，西临民心河两侧，东、北分别与万达 5A 级写字楼及万达公馆相邻。

Intercontinental Shijiazhuang Hotel is located in Yuhua District, close to Huaian Road. in the south and Minxin River in the west, adjacent to Wanda 5A office towers in the east and Wanda mansions in the north.

酒店西南角鸟瞰夜景

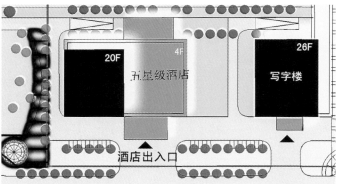

酒店总平面图

酒店立面夜景：简洁的线条、厚重的石材，充分展现代奢华酒店的高贵气质

酒店外立面

石家庄万达洲际酒店，为洲际集团的顶级品牌，超五星豪华酒店，房间豪华舒适。同时拥有面积不等的10间多功能厅，其中大宴会厅面积将近1300平方米。酒店设有餐厅、大堂吧、行政酒廊、健身中心等；为商务下榻、会议洽谈及休闲度假的理想之选。石家庄万达洲际酒店于2010年3月开工，于2011年9月23日开业。

Intercontinental Shijiazhuang is the top brand five-star hotel in Continental family with amazing deluxe rooms. This hotel has 10 multiple-function halls and the ball room among them is about 1300 m². There are restaurants, lobby lounge, exclusive wine bar, fitness center and other amenities for business travelling, conference, and vacation. Intercontinental Shijiazhuang started construction in March, 2010 and opened on September 23, 2011.

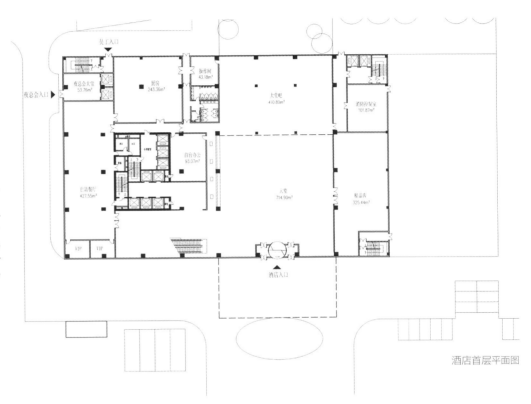

酒店首层平面图

酒店大堂

酒店景观　　酒店景观

酒店大堂

酒店套房

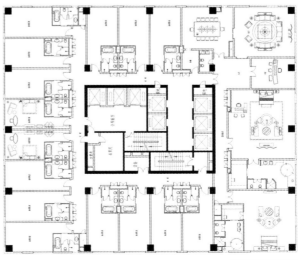

酒店塔楼平面图

酒店内装

大宴会厅

红酒吧

西安万达希尔顿酒店
HILTON XI'AN

西安万达希尔顿酒店位于西安市新城区民乐园地区，东二路以南、东新街以北、尚俭路以东和尚勤路以西。

Hilton Xi'an is located in Minle Park, Xincheng District, Xi'an, bordering Dong'er Road on its south, Dongxin Street to its north, Shangjian Road to its east and Shangqin Road to its west.

酒店用地面积 2.31 万平方米，总建筑面积 5.2 万平方米，其中地上 3.65 万平方米，地下 1.55 万平方米。地上 8 层，地下 1 层；地上部分均为酒店客房及配套功能，地下为机动车及非机动车车库以及酒店配套功能用房，室外设置完善的环境景观及配套系统。本酒店定位为五星级高档酒店。

The hotel's total site area is 23,100 m², with a gross floor area of 52,000 m², of which 36,500 m² is above ground and 15,500 m² is below ground. The above-ground eight floors occupy guest rooms and related supporting rooms, and the basement floor offers auto and non-auto car park and BOH rooms. Targeting to be an up-scale five-star hotel, it also provides well-designed landscape and facilities.

酒店内庭院景观夜景

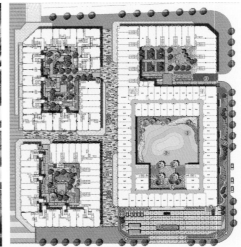

酒店总平面图

酒店外景

酒店入口

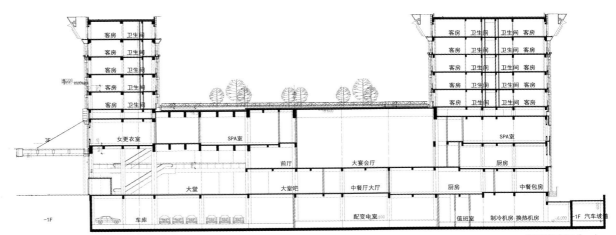

酒店剖面图（一）

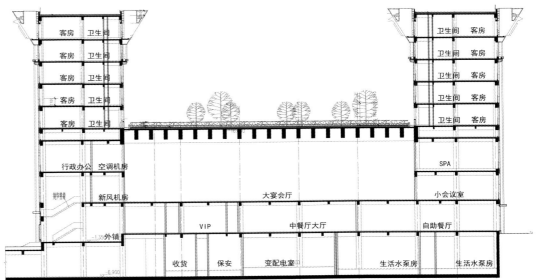

酒店剖面图（二）

酒店内庭院景观

酒店内庭院景观

酒店游泳池

酒店秉承五星级酒店特有的尊贵感，同时体现万达建筑一贯的简洁、现代、大气的设计风格，展现出企业独特的内在文化。立面设计汲取西安文化的精髓，以西安传统的文化元素融合到酒店设计细部中，同周边的其他建筑共同组成民乐园地区特色。西安万达希尔顿酒店于2009年12月开工，于2011年8月23日开业。

The hotel emanates prestige and luxury and also embodies Wanda's simple, modern and elegant style, representing Wanda's unique corporate culture. Inspired by the essence of Xi'an local culture, the facade incorporates local cultural elements into its detail design to blend in with the architectural context of the Minle Park. Hilton Xi'an started construction in December, 2009 and opened on August 23, 2011.

酒店大堂

酒店大堂

大宴会厅

总统套房

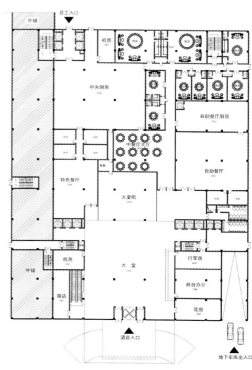

酒店首层平面图

常州万达
喜来登酒店
SHERATON CHANGZHOU XINBEI HOTEL

常州万达喜来登酒店，是常州新北万达广场城市综合体项目的一部分，位于常州市新北区，用地位于通江大道西侧，北侧隔内院与步行街、购物中心相望，西临步行街，南临太湖路。总建筑面积 3.74 万平方米，其中地上建筑面积 3.14 万平方米。

Sheraton Changzhou Xinbei Hotel is a part of Changzhou Xinbei Wanda Plaza Urban Complex Project. It is located in Xinbei District of Changzhou. The site is situated in the west of Tongjiang Road. Pedestrian streets and shopping centre could be seen through from the north of the site. It is adjacent to the pedestrian street to the west and Taihu Road to the south. The overall floor area is 37,400 m² and the total above-ground floor area is 31,400 m².

酒店外立面

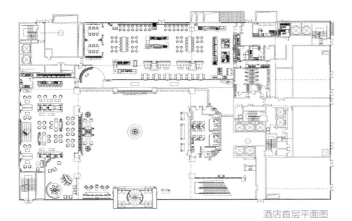

酒店首层平面图

酒店外立面

总客房数247间，设有大宴会厅1240平方米，大堂860平方米，行政酒廊240平方米，总统套350平方米。

It has a total of 247 guest rooms, a 1,240 m² banquet hall, a 860 m² lobby, a 240 m² administrative bar and a 350 m² presidential suite.

外立面设计以水为母题，通过玻璃幕墙折板角度的变化，写意水"倾流而下"。内装设计充分挖掘地方文化元素，具有典雅、大气的气息。

The exterior design is themed with water, using various angles of faceted glass curtain wall system, to create a scene of water flowing down. The interior design makes fully use of local culture elements, which is elegent and geneous.

景观设计营造静谧、高贵的氛围，并实现了与商业广场的有效分隔与联系。常州万达喜来登酒店于2010年7月20日开工、2011年12月10日开业。

The landscape design gives a sense of peace and nobility. At the same time, it makes a successful seperation and connection with the commercial plaza. Sheraton Changzhou Xinbei Hotel started construction in July 20, 2010 and opened on December 10, 2011.

酒店大堂

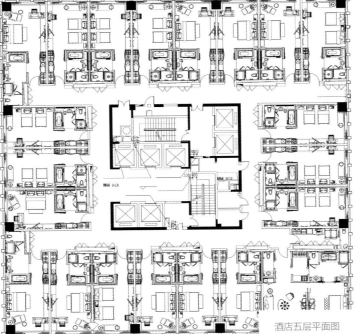
酒店五层平面图

总统套

全日餐厅

泰州万达希尔顿逸林酒店
DOUBLETREE BY HILTON TAIZHOU, JIANGSU

泰州万达希尔顿逸林酒店位于江苏省泰州市泰州万达广场，济川东路以北、海陵南路以西。酒店总建筑面积约4万平方米，拥有客房250间，于2011年12月9日开业。

Doubletree by Hilton Taizhou, Jiangsu is located in Taizhou Wanda Plaza, Taizhou City, Jiangsu Province. It situates to the north of Jichuandong Road, west to Hailingnan Road. The hotel with a total floor area of about 40,000 m², has 250 rooms, and it grandly opened on December 9, 2011.

酒店拥有完善的餐饮设施：拥有三家餐厅，包括一家中餐厅，一家特色餐厅和一家全日餐厅。此外还设有大堂吧、红酒吧。

The hotel owns perfect dining facilities: three restaurants including a Chinese restaurant, a restaurant with unique characteristic, and a all-day dining restaurant. In addition, it is equipped with lobby bars and wine bars.

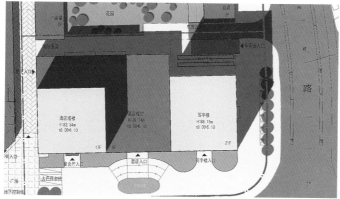

酒店总平面图

酒店主入口

酒店外立面

酒店外景

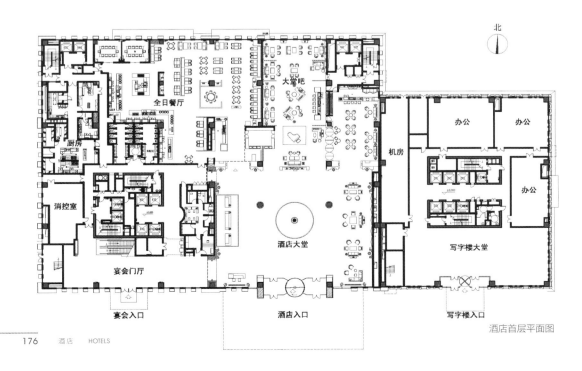

酒店首层平面图

酒店外立面夜景

酒店大堂：独具中国特色的空间

酒店的会议设施位于三层，包括一间无柱宴会厅（1200平方米）及5间会议室，一个商务中心。宴会厅可容纳800人，可分隔成三个部分使用。

Hotel's conference facilities are located on the third floor, which include a column-free banquet (1200 m²), 5 meeting rooms and one business centre. The banquet hall can hold 800 people, which can be sub-divided into three rooms for different usage.

酒店的康体设施位于四层，包括恒温泳池、健身中心、美发沙龙。

Hotel's recreation facilities are located on the fourth floor which include swimming pools, fitness centers and hair salons.

酒店拥有客房标准间235间，普通套房15套，部长套房2套，总统套房1套，酒店北层设有行政酒廊。酒店客房标准间面积约为42平方米。

The hotel has 235 standard guest rooms, 15 units of ordinary suites, 2 units of minister suites and one president suite. The executive lounge is on the sixteen floor. The floor area of hotel's standard guest room is around 42 m².

总统套房

红酒吧

镇江万达喜来登酒店
SHERATON ZHENJIANG HOTEL

镇江万达喜来登酒店拥有289间装饰精美典雅的客房与套房；盛宴全日餐厅、采悦轩中餐厅、"雅"日餐厅及红酒吧、大堂酒廊等餐饮设施；超过1650平方米的宴会与会议场地，包括1个1200平方米的无柱式大宴会厅、4个多功能厅和1间贵宾厅，均配备高级试听设备，可满足大型会议和高档商务接待等不同需求；除此之外，酒店的休闲设施包括喜来登健身中心和室内泳池等休闲设施。

Sheraton Zhenjiang Hotel possesses 289 well-decorated guest rooms and suites, food services such as Shengyan All Day Dinning Restaurant, Caiyuexuan Chinese Restaurant, "Ya" Japanese Restaurant, red wine bar, and a lobby bar. It also owns a more than 1,650 m² banquet and meeting area, including a 1,200 m² astylar banquet hall, 4 multifunctional hall and a VIP hall, which are all equipped with advanced audition facilities to meet various demand such as big conference and top business reception. In addition, its leisure facilities contain Sheraton fitness centre and indoor swimming pool.

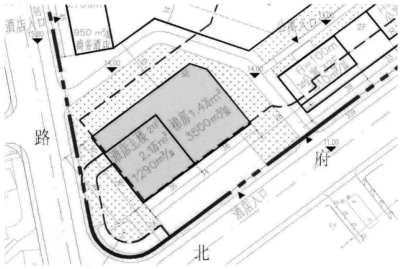

酒店总平面图

酒店日景

镇江万达喜来登酒店采用浅米黄色石材幕墙，配以金属装饰，造型庄重典雅，加上丰富多彩的景观、时尚高雅的内部装修，一起构成镇江最豪华的五星级酒店。镇江万达喜来登酒店于2009年12月29日开工，于2011年8月12日开业。

Sheraton Zhenjiang Hotel uses light cream-coloured stone wall with metal ornaments. Being solemnly and delicately shaped, with plenty of colourful landscapes and fashionable and delicate interior decoration, all of these consititute the most luxury five-star hotel in Zhenjiang. Sheraton Zhenjiang Hotel started construction on December 29, 2009 and opened on August 12, 2011.

酒店外立面

酒店入口

酒店夜景

酒店首层平面图

酒店大堂

宴会厅

大堂吧

总统套房卧室

廊坊万达
希尔顿逸林酒店
DOUBLETREE BY HILTON LANGFANG

廊坊万达希尔顿逸林五星级酒店是廊坊万达广场建筑群的重要组成部分，酒店地上总建筑面积约 3.6 万平方米，是廊坊市目前最高星级的酒店。房间奢华舒适，同时拥有面积不等的多间多功能会议厅，大宴会厅面积将近 1200 平方米。酒店设有全日餐厅、中餐厅、西餐厅、酒吧、行政酒廊、健身中心、泳池等配套设施。酒店于 2010 年 10 月开工，于 2011 年 11 月开业。

Doubletree by Hilton Langfang forms an integral part to the Langfang Wanda Plaza. With an above-ground gross floor area of 36,000 m², the hotel stands as the tallest star hotel in Langfang. It offers luxury and comfortable rooms, multi-function rooms of different sizes and a 1,200 m² ballroom. The hotel also offers all-day dining restaurant, Chinese restaurant, western restaurant, pub, executive, gym, swimming pool and other facilities. The hotel stanted to construct on October, 2010, and was opened on November, 2011.

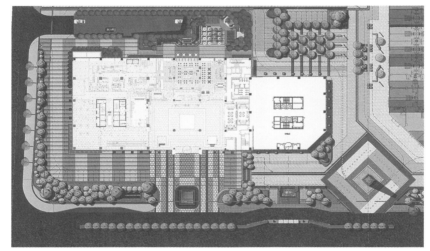

酒店总平面图

酒店外立面

酒店外立面

酒店外立面

酒店门头

酒店外立面

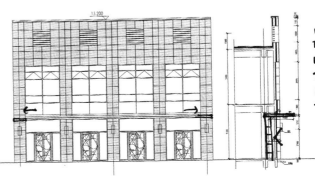

大师草图

WHEN WE NEED TO RECESS AT 1 FL, WE USE THE CANOPY TO DO SO. WE EXTEND ALSO THE CANOPY ON BOTH SIDES —

酒店大堂

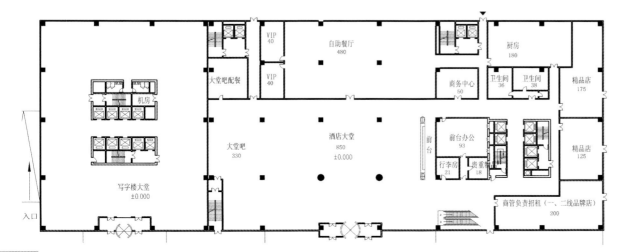
酒店首层平面图

宴会厅

总统套房

VIP会议室

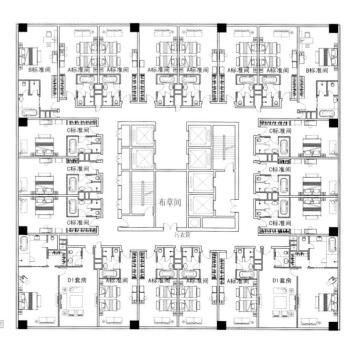

酒店标准层平面图

大庆万达
喜来登酒店
SHERATON DAQING HOTEL

大庆万达喜来登酒店坐落在世纪大道北侧，规划一路东侧的位置，具体地址为：经二街 2 号。酒店建筑面积 43500 平方米，占地面积 3900 平方米。酒店于 2010 年 3 月开工，2011 年 11 月 25 日盛大开业，是目前大庆市唯一的国际品牌的 5 星级酒店。 毗邻机场高速公路，25 分钟可轻松抵达大庆萨尔图机场（DQA）。 与酒店相连的大庆万达广场拥有众多的购物商铺、娱乐场所、写字楼和公寓。

Sheraton Daqing Hotel is located in the north side of Century Avenue and in the eastern side of No.1 Guihua Road. Its location address is No.2 Jing'er street. The hotel's GFA is 43,500 m^2, which covers the total site area of 3,900 m^2. It grandly opened on November 25, 2011, which is the only international brand five-Star Hotel in Daqing. Adjacent to the airport expressway, it just takes 25 minutes from hotel to Daqing Saertu Airport (DQA). Daqing Wanda Plaza connected with the hotel has many shops, entertainments, office buildings and apartments.

酒店有 290 间宽敞舒适的客房及套房，可让客人尽享轻松惬意的温馨下榻体验。每间房内均设有按人体脊椎学理论设计的喜来登专属"甜梦之床™"、37 英寸 LCD 纯平电视和热带雨林淋浴间。酒店拥有 1200 平方米的大宴会厅和另外 50～160 平方米 6 间现代化多功能厅，均配有一流的会议和音响设施，定会成为举办商务会议活动和社交聚会的理想之地。大庆万达喜来登酒店，商务，度假两相宜。

The hotel has 290 spacious and comfortable guest rooms and suites, which let guest enjoy comfortable travel experience. "Sheraton Sweet Sleeper Bed"— according to the human spine theory is provided in each room. Furthermore, it offers exclusive 37 inch LCD flat screen TV and tropical rain forest shower. The hotel has a large banquet hall of 1,200 m2 and 6 modern multi-functional halls from 50 to 60 m2, equipped with first-class conference and sound equipment, which will be a ideal place to hold business meetings and social gathering events. Daqing Wanda Sheraton Hotel properly combines business with leisure.

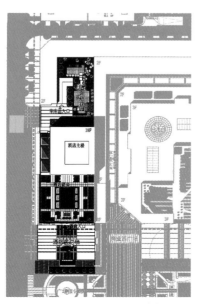

酒店总平面图

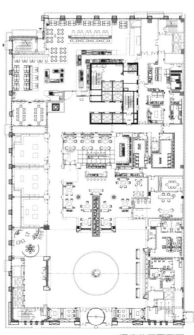

酒店首层平面图

酒店入口

酒店大堂　　　　　　　　　　　　　　　会议厅　　　　　　　　　　　　　　　酒店套房

PART 3 万达学院
WANDA INSTITUTE

万达学院
WANDA INSTITUTE

万达学院是万达集团的培训基地,是国内顶级的企业文化学院之一,也是国内首家获得绿色建筑三星设计标识认证的学校类建筑。

Wanda Institute, the training base of Wanda Group, is one of the top enterprise colleges in China and is awarded to be the first three-star green building for institutional architecture.

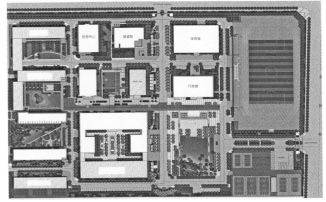

总平面图

纯净的天空与纯粹的建筑相映成趣

教学楼夜景

万达学院园区内规划有行政楼、教学楼等主要建筑,同时还设有体育馆、餐厅、宿舍、展览馆等相对附属的建筑。园区的规划主次有序、分区明确,体现了开朗大气的规划主旨和高效严谨的企业特质,既表现了百年万达的企业文化气质,又具备了欧式高等学府的风范。

The campus is planned to have main buildings including administrative building, teaching building, and other components including gymnasium, dining hall, dormitory, exhibition hall and so on. The master plan of campus has very clear logic of programming and zoning, which is good to express the planning's openness and grandeur, also the enterprise's features of high efficiency and strictness. Generally, this campus demonstrates Wanda's enterprise spirit and the demeanour of European colleges very well.

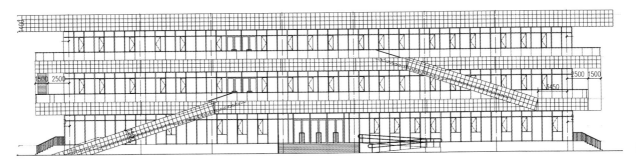

剖面图

流动的光影、充满张力的建筑外形，赋予教学楼浓厚的韵律感

夜幕中柔和静谧的行政楼

万达学院位于河北省廊坊市花园道与梨园路交叉口西南侧，项目占地约 13.3 公顷，总建筑面积约 9.9 万平方米，其中一期建筑面积 5.2 万平方米，二期建筑面积 4.7 万平方米。

Wanda Institute, located in the southwest of intersection of Huayuan Road. and Liyuan Road. in Langfang City of Hebei Province, covers a site of 133,000 m². And the gross floor area is about 99,000 m², including phase one of 52,000 m², phase two of 47,000 m².

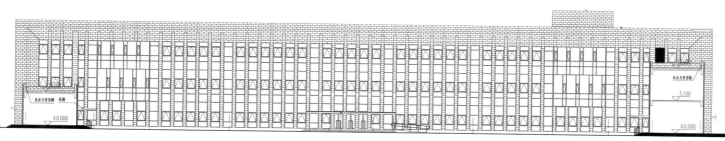

教学楼立面图

简洁干练的教学楼入口空间

柔和的灯光中,教学楼的柱廊倒映在水面上,静谧悠长

建筑与雕塑、流水与灯光,晚风习习,带来芳香,这是一天中最美好的时光

粗犷的廊架为展览馆套上了精美的画框

万达学院体育馆门厅:简洁、舒适的多功能运动空间

万达学院以其简洁震撼的立面设计、典雅的室内设计、幽静的景观设计、浪漫的夜景设计、人文的导向设计为主线,各专项设计内外呼应、浑然一体,勾勒出万达学院美轮美奂的建筑画卷。

Wanda Institute has concise and apealing façade, elegant interior, peaceful landscape, romantic nightscape and humane way finding system, which work together to create one stunning architecture picture scroll.

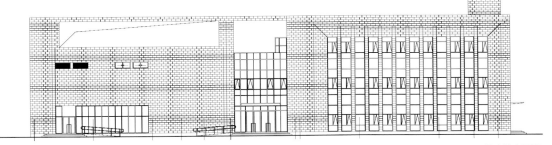

行政楼立面图

万达学院餐厅

万达学院图书馆:弧形的书架,舒适的沙发使阅读变得很惬意

万达学院 KTV

万达学院报告厅:墙面木制编织的视觉效果同时满足了声学功能

万达学院游泳池

万达学院教学楼首层　　　　　　　　　　　　　　　　　　　　　　　　万达学院行政楼门厅

万达学院餐厅　　　　　　　　　　　　　　　万达学院篮球馆

PART 4 景观
LANDSCAPE

万达商业广场室外景观是向城市开放、同城市相互融合的公共空间。规划设计强调商业突出、交通优先，形成能举办大型商业活动的集散广场，与城市交通形成港湾式的接驳点。景观设计通过设置现代时尚的景观小品及雕塑来活跃广场商业氛围。主要的核心景观元素包括水景、雕塑、LOGO 塔、导视牌、树阵、灯箱、铺装、坐凳等。

The outdoor landscape of Wanda Plaza embraces the urban public space and seamlessly merges into the city texture. The site planning focuses on commercial atmosphere and circulations to create a gathering place holding major commercial activities and connecting urban public transportation. The landscape design employs modern cutting-edge fixtures and sculptures to activate the outdoor commercial environment. Key landscape elements include water feature, sculptures, logo structure, wayfinding system, tree grove, lighting fixture, pavement and outdoor seating etc.

万达酒店景观包括酒店前场景观及酒店后花园景观两部分，强调舒适性、私密性和奢华感。景观元素体现出高度的艺术感和观赏性，营造出高星级酒店的奢华度假氛围。核心景观元素包括主入口水景、后花园水景、雕塑、景观大树和组团式植物群落等。

Wanda Hotel landscape strives to deliver the sense of comfort, privacy and luxury through carefully configured front arrival and back amenity area. Applied artful and ornamental landscape elements actively fabricate a high-end resort feeling. Key landscape elements consist of primary arrival water element, amenity area garden water feature, sculpture, specimen tree and structured vegetation massing.

广场雕塑

广场雕塑

广场雕塑

广场雕塑

广场景观　　　　　　　　　　　广场雕塑

广场雕塑

酒店景观

酒店景观

广场景观　　　　　　　　　　　　　广场雕塑　　　　　　　　　　　　　广场景观

学院雕塑

景观雕塑

广场雕塑

广场景观

广场景观

酒店景观

酒店景观

广场景观

广场景观

广场景观

PART 5 项目索引
INDEX OF THE PROJECTS

项目索引 | 2011
INDEX OF THE PROJECTS 2011

万达广场 / WANDA PLAZAS

唐山路南万达广场
TANGSHAN LUNAN WANDA PLAZA
2011.12

常州新北万达广场
CHANGZHOU XINBEI WANDA PLAZA
2011.12

泰州万达广场
TAIZHOU WANDA PLAZA
2011.12

福州仓山万达广场
FUZHOU CANGSHAN WANDA PLAZA
2011.12

大庆萨尔图万达广场
DAQING SAERTU WANDA PLAZA
2011.11

廊坊万达广场
LANGFANG WANDA PLAZA
2011.11

郑州中原万达广场
ZHENGZHOU ZHONGYUAN WANDA PLAZA
2011.10

石家庄裕华万达广场
SHIJIAZHUANG YUHUA WANDA PLAZA
2011.09

银川金凤万达广场
YINCHUAN JINFENG WANDA PLAZA
2011.09

厦门湖里万达广场
XIAMEN HULI WANDA PLAZA
2011.09

武汉经开万达广场
WUHAN JINGKAI WANDA PLAZA
2011.08

镇江万达广场
ZHENJIANG WANDA PLAZA
2011.08

上海江桥万达广场
SHANGHAI JIANGQIAO WANDA PLAZA
2011.06

淮安万达广场
HUAIAN WANDA PLAZA
2011.01

酒店/ HOTELS

唐山万达洲际酒店
INTERCONTINENTAL TANGSHAN
2011.12

常州万达喜来登酒店
SHERATON CHANGZHOU XINBEI HOTEL
2011.12

泰州万达希尔顿逸林酒店
DOUBLETREE BY HILTON TAIZHOU, JIANGSU
2011.12

南京万达希尔顿酒店
HILTON NANJING
2011.11

廊坊万达希尔顿逸林酒店
DOUBLETREE BY HILTON LANGFANG
2011.11

济南万达凯悦酒店
HYATT REGENCY JINAN
2011.11

大庆万达喜来登酒店
SHERATON DAQING HOTEL
2011.11

石家庄万达洲际酒店
INTERCONTINENTAL SHIJIAZHUANG
2011.09

西安万达希尔顿酒店
HILTON XI'AN
2011.08

镇江万达喜来登酒店
SHERATON ZHENJIANG HOTEL
2011.08

广州白云万达希尔顿酒店
HILTON GUANGZHOU BAIYUN
2011.08

武汉万达威斯汀酒店
The WESTIN WUHAN WUCHANG
2011.07

万达学院 / WANDA INSTITUTE

万达学院
WANDA INSTITUTE
2012.02

楚河汉街 / CHU RIVER & HAN STREET

武汉中央文化区楚河汉街
Wuhan Central Cultural Tourism District
Chu River & Han Street
2011.09

万达学院
WANDA INSTITUTE

万达学院行政楼

创立万达学院是万达集团为实现"百年企业"战略目标，解决人才短板而采取的新的管理举措，在王健林董事长的亲切关怀下，历时1年半，耗资7个亿打造而成。万达学院的建成将填补中国商业地产和旅游地产行业理论的空白，将成为中国超一流的世界级品牌文化旅游度假区的理论倡导者、创新者和实践者。

Wanda Institute was established to fulfill Wanda Group's strategic objective of becoming a "century-old Enterprise", as a new initiative to specifically solve the problem of talents shortage. It was crafted with 700 million Yuan, within a year and a half and under the care of Chairman Wang Jianlin. The establishment of Wanda Institute fills the blanks in the industrial theories of China's commercial and tourism real estate. It will become a theory advocate, innovator as well as practitioner of China's top-class cultural tourism resorts brand.

学院概况
Overview

2012年，万达集团的员工总数达到七万余人，每年新招聘的员工数量预计超过一万人。在业务高速发展的今天，集团对于人才培养的要求越来越高、越来越迫切。2009年的年会上，董事长提出"万达学院非办不可！"万达学院主要针对万达集团内部经理级以上管理者，每人每年在学院的平均受训时间约为5天。万达学院每年培训超过1万人次。

The staff of Wanda Group will reach 50,000 in 2012 and new staff recruited each year are expected to exceed 10,000. With its rapid business growth, the Group's requirement on the training of talents is becoming increasingly high and urgent. In the Group's 2009 Annual Summit, the Chairman proposed that "the establishment of Wanda Institute is a must!" Staff of managers within the Group and they shall be trained at Wanda Institute for around 5 days each year, in average. Total persons attending trainings at Wanda Institute each year exceeds 10,000.

学院落户在廊坊高新经济开发区，占地13.3公顷，距离北京CBD总部仅40公里。一期建筑面积5.22万平方米，投资7亿元人民币。由行政楼、教学楼、公寓楼、餐饮楼、室内体育馆、室外运动场、信息中心等建筑组成。花园式的院区设计，并配备国内一流的教学服务设施，目前，可同时满足700多人同时在学院培训。

Wanda Institute is located in the High-tech Economic Development Zone of Langfang City, occupying an area of 200 mu (approximately 133,000 square meters), only 40 kilometers away from Beijing's Central Business District. Phase I of Wanda Institute has a floor space of 52,200 square meters, with an investment of 700 million Yuan. It consist of an administration building, a teaching building, an apartment building, a dining hall, an indoor gymnasium, an outdoor stadium, an information center and other buildings. Design of the campus adopts the garden style and equips with China's best teaching service facilities. Currently over 700 people can be trained at Wanda Institute at the same time.

学院特色
Features of the Institute

以"有用"为教学理念
"Usefulness" as the Teaching Concept

学院以"有用"为校训；以弘扬万达企业文化、沉淀传播专业知识、推动组织关系改善、健全员工职业心态为己任；汇聚、调动1万名万达各级管理人员的集体智慧和能量，帮助企业提升效率。

The Institute uses "usefulness" as its motto; it tasks itself with advocating Wanda's corporate culture, consolidating and spreading expertise, promoting the improvement in organizational relations and refining the professional mentality of the staff; it converges and mobilizes the collective wisdom and energy of 10,000 Wanda managers and helps the Group upgrade its efficiency.

通过复盘机制建立组织防御体系
Establishing the Organizational Defense System through the Readjustment Mechanism

为发现公司真实存在的问题，万达学院把组织诊断作为一项重要的工作方向。董事长指出，"复盘工作既是研究创新，也是管理推动"。万达做商业地产就是一个做项目管理的过程，其中会涉及多个部门和专业。每个项目做完，万达学院就会对其进行复盘分析，通过一种结构化的方法系统性地发现组织中真实存在的问题，比如流程问题、制度问题、个人问题。每个问题都有可能成为课堂培训研讨的专题。通过不断的项目复盘，不断解决业务问题，将企业风险降到最低。

To discover real problems in the company, Wanda Institute regards organizational diagnosis as an important working direction. Chairman Wang pointed out that "the readjustment work is both a research innovation and a management promotion". Wanda's commercial real estate business is a process of project management, which will involve multiple departments and specialties. After completion of each project, Wanda Institute will conduct readjustment analyses, utilizing a structured method to systematically discover the real problems in the organization, e.g., problems with the process, system and the staff members themselves. Each of these problems can become a special topic in the seminar. Through constant project readjustment, business problems, and the risks to the enterprise will be lowered to the minimal level.

着力打造万达特色的培训工具
Focusing on Crafting Training Tools with Wanda's Characteristics

万达学院培训主要关注三类问题：集团问题、部门问题、个人问题。针对集团问题，"复盘法"就是从集团全局关系层面进行问题诊断的一种模式。对于部门问题，紧盯年度任务，梳理抱怨最多、影响最大的问题，并将之列为培训重点。对于个人问题，采用"能量集市"法，是一种群体帮助个人解决问题的教学方法。

Wanda Institute's training focuses on three categories of problems: the problems with the Group, the problems with the departments and the problems with the staff members themselves. For the problems with the Group, The "readjustment method" is a model to diagnose problems on the level of the Group as a whole. For problems with the departments, the Institute focuses on the annual tasks, concentrates on the solutions to the problems with most complaints and adverse influences and lists them as the emphasis of the trainings. For problems with the staff members themselves, the "energy fair" approach is adopted, which is a training method for the collective to help individuals solve their own problems.

搭建集团知识管理平台
Building the Group's Knowledge Management Platform

以组织诊断为出发点，通过教学活动和教学管理工作，牵头集团层面的各类教材开发，收集学员问题和学员案例，充分利用学院平台和网络平台，输出课题研究成果和锦囊库，最终达到"组织有问题找学院，学员有问题找学院"。

Wanda Institute starts from organizational diagnosis and leads the development of various types of teaching materials through teaching activities and teaching management. It collects questions and case studies from students and fully utilizes the Institute's platform and Internet platform to export topic research results and "bank of tips". Finally, "when the organization has problems, it comes to the Institute; while when the students have problems, they come to the Institute as well."

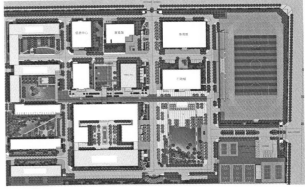

总平面图

万达商业规划研究院文化旅游分院
THE CULTURAL TOURISM BRANCH OF WANDA COMMERIAL PLANNING & RESEARCH INSTITUTE

万达商业规划研究院文化旅游分院

万达商业规划研究院注册成立于2007年，是全国唯一一家从事商业及文化旅游项目规划设计（同时进行全过程管控）的技术管理和研究机构，是万达集团的技术归口管理部门。

Wanda Commercial Planning & Research Institute Co., Ltd. was established in 2007, as China's only technical management and research institution engaged in the planning and designing of commercial and cultural tourism projects at the same time of conducting whole-course control. It is Wanda Group's competent department for technical management.

文化旅游分院隶属于万达商业规划研究院，是目前国内专门从事文化旅游项目设计、管控及研发的综合性机构，具有承接国际、国内大型文化旅游项目的能力，主要负责文化旅游项目整体规划、设计、市政、景观配套，涵盖主题公园、秀场、滑雪场、高尔夫球场、影视基地等内容，覆盖北京、广州、大连、长白山、西双版纳和莫斯科、孟买等国内外多个城市或地区。

One part of the Institute is its Cultural Tourism Branch, which is a versatile organization dedicated to the designing, controlling and R&D of Wanda Group's cultural tourism projects. It is capable of undertaking international and domestic large-scale cultural tourism projects. The Cultural Tourism Branch is mainly responsible for the master planning, designing, municipal planning and landscaping supporting facilities of projects including theme parks, show theatres, ski resorts, golf courses and movie & TV bases. Geographically, it covers numerous domestic and foreign cities or regions including Beijing, Guangzhou, Dalian, Changbaishan, Xishuangbanna as well as Moscow and Mumbai.

万达商业规划研究院文化旅游分院下设管理所、规划所、建筑所、剧场所、室内主题公园所、室外主题公园所、剧场设备所、游乐设备所、高尔夫所共九个专业部所。现有员工近120人，均为兼具设计与管理经验的复合型人才，其中：研究生以上学历者占员工总数的50%；海外留学归国人员占员工总数的12%，精通英、德、俄、印度等多国语言；拥有各专业注册执业资格者占员工总数的33%，拥有高级技术职称者占员工总数的25%，涉及城市规划、市政道桥、建筑设计、环境设计、特种设备、舞台效果等多个专业领域。

The Cultural Tourism Branch has 9 departments: the Management Department, Planning Department, Theatre Department, Indoor Theme Park Department, Outdoor Theme Park Department, Theatre Equipment Department, Amusement Equipment Department and Golf Department. Its staff is close to 120, all versatile talents with both designing and management experience. Among them, those with Master's Degree and above account for 50% of all. 12% of them have degrees of overseas universities and are proficient in English, German, Russian and languages of numerous foreign countries like India. 33% of them have qualifications as registered practitioners of various specialties and 25% of them hold senior technical titles, involving numerous professional fields including urban planning, municipal roads and bridges, architectural designing, environmental designing, special equipment and stage effects.

万达商业规划研究院文化旅游分院目前正在进行多个国际级文化旅游项目的设计与管控工作，包括长白山国际度假区、西双版纳国际度假区、大连金石国际度假区和武汉中央文化区等，总投资超过千亿人民币。万达商业规划研究院文化旅游分院将继续致力于引领文化旅游规划发展，不断提升管理和技术水平，力争成为具有国际影响力的专业机构。

The Cultural Tourism Branch of Wanda Commercial Planning & Research Institute is currently undertaking the designing and controlling of multiple international-level cultural tourism projects, including Changbaishan International Resort, Xishuangbanna International Resort, Dalian Jinshi International Resort and Wuhan Central Cultural District, with a total investment exceeding 1,000 billion Yuan. The Cultural Tourism Branch of Wanda Commercial Planning & Research Institute will continue engaging itself in leading the development of cultural tourism planning and ceaselessly upgrade its management and technical capabilities, striving to become a specialized institution with international influence.

曾明	范纯松	齐宗新
郭峤宇	杨艳坤	潘立影
贺荣梅	邵汀潇	王惟
刘钊	邓金坷	张涛
戎帅	李甜	郭薇
李海龙	李楠	张帆
安竞	汪淼	李昕
	孟祥宾	宋樱樱
	石紫光	刘冰
	张宝鹏	牛晋华
		程鹏
		李为状
		王明妍

		青云富
		李文娟
		谢绪锦
龙向东		天迎斌
尚海燕		李杨
刘佳		唐海江
程欢	屠波	康斌
何海川	叶甲刚	谷建芳
秦好刚	张洋	刘平
郑斐	刘子瑜	孙楠
韩茂俊	徐小莉	雷磊
张琳	王燕	刘婷
李广	王魏巍	江勇

赖建燕	孙培宇	谭耀辉	王群华	李晴
黄大卫	莫力生	文善平	吴昊	王雪松
朱其玮	杨成德	岳勇	赵传建	谢寰
叶宇峰	黄引达	杨旭	李斌	杨宜良
王元	魏建彤	田杰	谷全	沈余
王绍合	范珑	黄勇	朱莹洁	罗沁
冯腾飞	李峻	张飚	王鑫	陆峰
刘冰	梅咏	刘宾	毛晓虎	王宇
吕圣龙	袁志浩	陈海燕	郝宁克	高加卿
康军	李峥	李浩然	阎红伟	周澄
国文	马红	张涛	万志斌	李江涛
孟宪民	马家顺	吴昊	陈杰	钟光辉
吕永军	付恒生	侯卫华	谷强	

2011
万达商业规划研究院
WANDA COMMERCIAL PLANNING & RESEARCH INSTITUTE CO., LTD

侯宝永	朱晓辉	李小强	黄春林	熊伟
张宁	刘佩	张立峰	翟锡葵	苗凯峰
张宇	高昊	赵鑫	王弘成	刘悦飞
王家林	刘江	刘辰宇	吴绿野	李彬
黄国辉	魏成刚	刘杰	兰峻文	史洪伟
李跃刚	李兵	党恩	温亚玲	鲍力
秦彬	李树靖	高振江	张振宇	王巍
谢剑	吴迪	马升东	杨洪海	孙佳宁
郑海	杨世杰	孙海龙	王朕	曹莹
耿大治	张振宇	古云	屈娜	莫鑫
刘阳	徐立军	王力平	卢波	王光宇
石路也	杨威	杨伍拾	孙辉	任永志
章宇峰	张后胜	杨云森	董丽梅	蓝毅

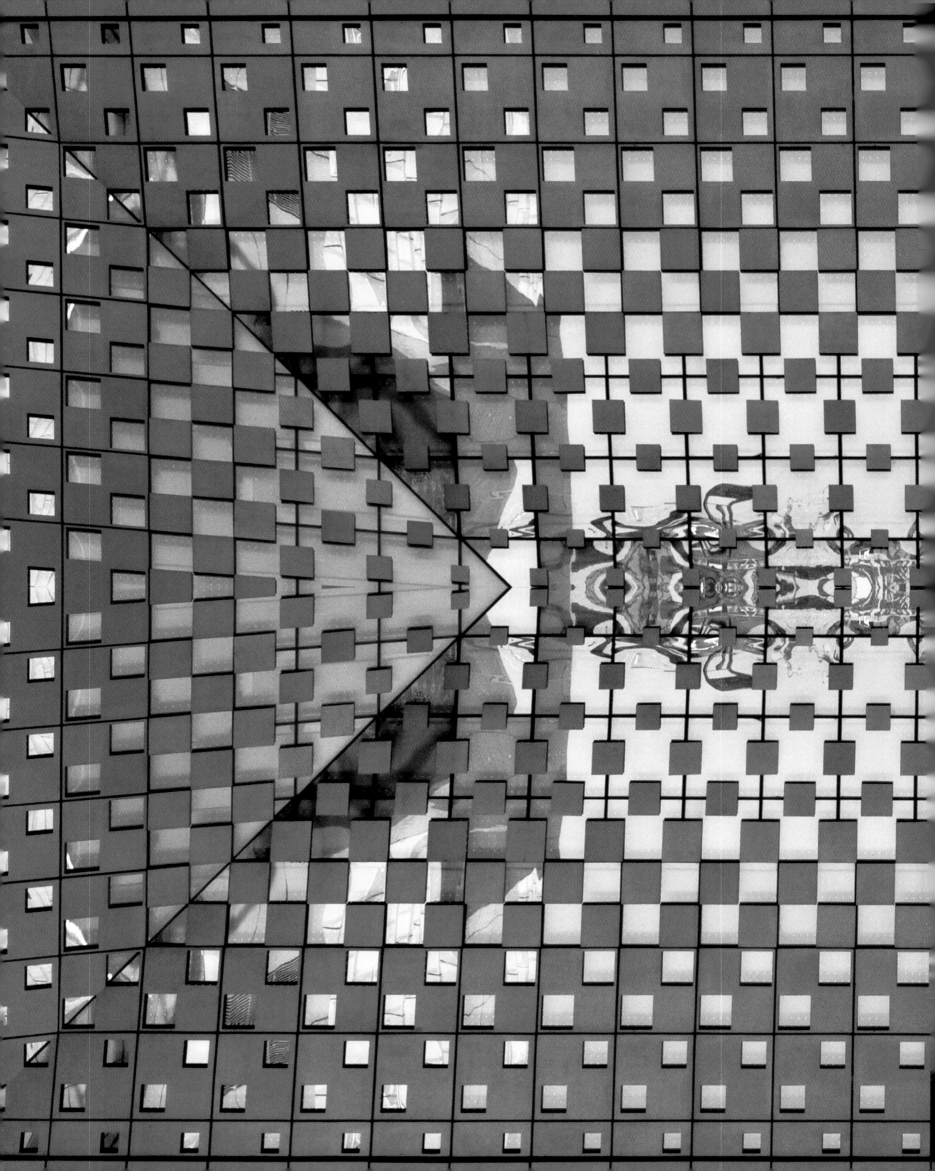

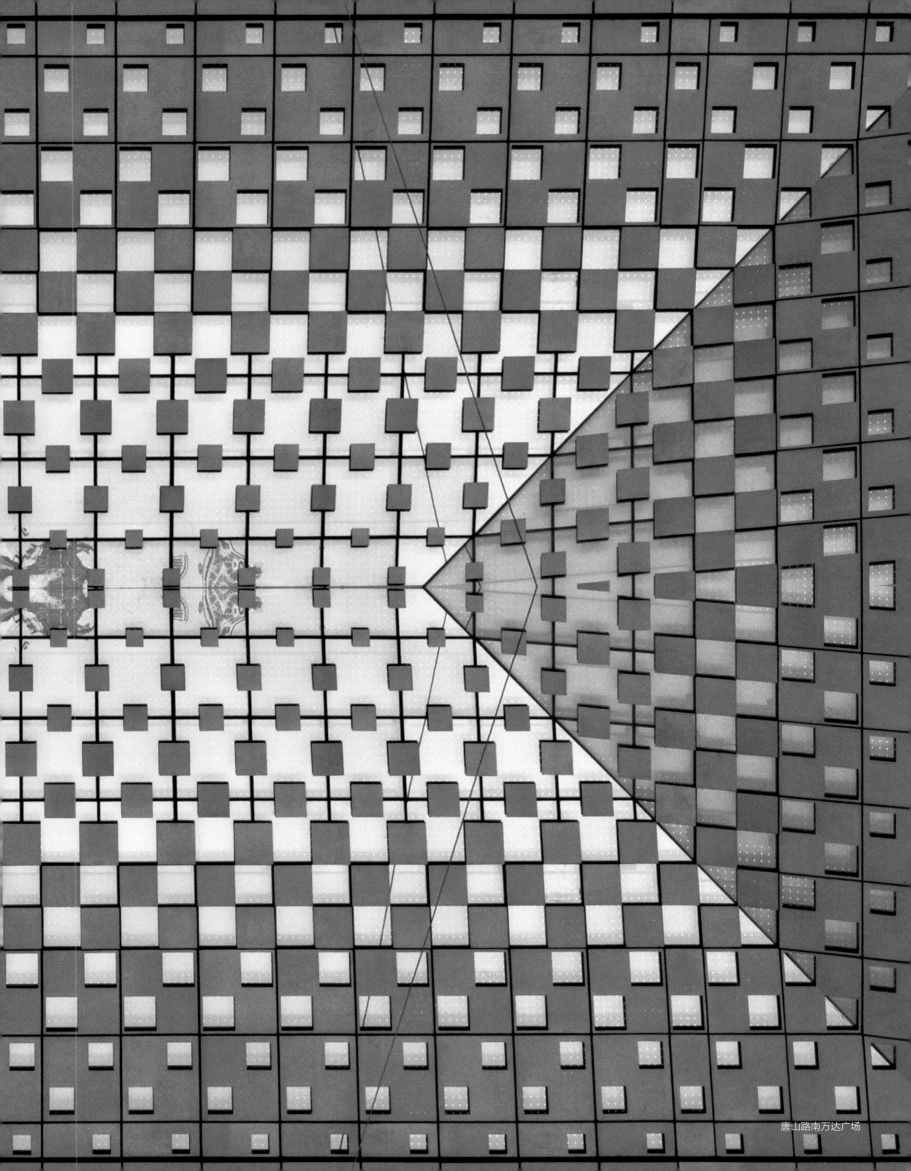
唐山路南万达广场